D0278720

Work design in practice

Work design in practice

Edited by
Christine M. Haslegrave
John R. Wilson, E. Nigel Corlett
University of Nottingham, England
and Ilija Manenica
University of Split, Yugoslavia

Proceedings of the Third International Occupational Ergonomics Symposium, Zadar, Yugoslavia, 18–20 April 1989

Taylor & Francis
London • New York • Philadelphia
1990

UK Taylor & Francis Ltd, 4 John Street, London WC1N 2ET

USA Taylor & Francis Inc., 1900 Frost Road, Suite 101, Bristol, PA 19007

British Library Cataloguing in Publication Data

Work design in practice.
1. Ergonomics
I. Haslegrave, C. M.
620.8′2

ISBN 0-85066-770-4

Library of Congress Cataloging-in-Publication Data is available on request

Typeset by Electronic Village Limited, Richmond, Surrey

Printed in Great Britain by Burgess Science Press, Basingstoke, Hants

Contents

Preface

This is the third in the series of International Occupational Ergonomics Symposia at Zadar in Yugoslavia, each of which explores a separate theme within ergonomics. The present theme of *Work Design in Practice* was chosen to look at new developments in the field and to identify new research directions, as well as to discuss the practical implementation of work design initiatives. Although work design can relate to a particular aspect of a task such as posture or equipment design, a major interest was in looking at work as a total experience and in seeing how this influences a person's motivation and acceptance of the work situation. We hoped to generalise from the various experiences and to provide an overview which would lead to more effective implementation.

The range of papers shows how widely practitioners interpret the scope of work design, and the programme tried to reflect the different perspectives. It spanned workplace design, job design, work organisation, future workplaces and involvement of workers in work changes, dealing with different styles of ergonomic evaluation and improvement. The papers raised the issues we are facing as practitioners and demonstrated to some extent the influence of culture, with speakers from so many different countries.

Many basic concerns of ergonomists were discussed: How can a designer know whether the designed workplace is good or not? Where does the designer obtain feedback? Work design frequently means work redesign, and one of the major current issues is the difficulty in identifying the workplace changes which are necessary. We are still very much at the stage of attacking individual problems, and cannot yet see a methodology emerging. However, it can be seen from many of the papers that the working environment plays a large part in the problem, with contributions from social and organisational issues and the influence of the management ethos. There was clear recognition that practitioners must work with the people concerned with the changes—it is not sufficient to present a plan to management alone. We need to justify the effectiveness of changes to both operators and management, and to involve both in identifying and making the changes, but we also need the support of engineers and designers in order to be able to make changes in the workplace. We need support and effort from all of these groups to be sure that the changes will be implemented and continue to be effective.

Zadar's friendly, relaxing atmosphere and warm hospitality are ideal for the exchange of views and experiences among those working in the field, and we should like to thank all our colleagues at the Faculty of Science and Arts there for their help in organising the Symposium. Our thanks also go to Christine Stapleton and Maria Reyes for compiling the indexes, and to Robin Mellors of Taylor and Francis for his patience and advice in the editing.

Nottingham, 1990

Christine Haslegrave
John Wilson
Nigel Corlett
Ilija Manenica

Part I

Introduction and overview

Chapter 1
An integrated approach to the design of work

E. Nigel Corlett

Department of Production Engineering & Production Management, University of Nottingham, Nottingham NG7 2RD, U.K.

To many it may seem odd that it is useful to present a paper on 'an integrated approach to the design of work'. It is nearly fifty years since the appearance of the Tavistock publications with their ideas for new working practices and their arguments for them, which they christened 'socio-technical systems theory'. Since then developments, explorations and applications of ideas arising from their work, and the reappraisal of methods for work organisation from earlier days, have led to a plethora of case studies, textbooks, investigations, visits, conferences and experiments across the industrialised and industrialising countries.

So, is there anything left to say about the subject? Probably, quite a lot! There are still many questions unanswered. Why have the underlying ideas found better soil in some countries than others? Why is so much attention paid to job organisation and content and so little to basic ergonomics? Why are ethics so little discussed, as if we were ashamed to recognise any higher purpose than running a profitable business, and why are we only now, with a few notable exceptions, beginning to look at the costs and benefits at a less trivial level than the direct costs of change and the immediate commercial gains from increased output and decreased turnover?

In many parts of the Western world, and not least in Britain, the myth of the 'practical man' as the epitome of the industrial manager has grown again. Managing people, it is assumed, is a matter of telling them what has to be done, watching that they do it and punishing them if they do not. This is a slight caricature, but very recognisable in its style. Understanding management by understanding the views and desires of people has been on the retreat whilst the mythology flourishes. There will be a change. It is already visible as a result of a shortage of people willing to undertake the arduous training for industry, the increased expectations of working people and their resentments at having increased restrictions on their lives. The political pendulum will swing to a more equitable society and the recognition of the part played by work experiences in the context of the shaping of society will become more visible. Once again we shall be on a rising tide, to carry us on the way to improvements in society and in the quality of all our lives.

From a European perspective, the programme of research sponsored by the Dublin-based European Foundation for the Improvement of Living and Working Conditions (1989) includes major investigations over the next four years into job design and the social impacts of work on life. These are encouraging pointers to the future, and an international indication of the political recognition of the importance of work in society apart from its economic contribution.

However, it is the experiences people have in society which are major determinants of whether or not work designs are seen as appropriate, obvious and natural, or as the manifestation of some eccentricity. It is no accident that Scandinavia, and particularly Sweden, has been the major centre for practical applications of work design ideas, due to its social and political attitudes to consensus. The uphill struggle in the U.S.A. and in the U.K. comes from deep-seated views of the opposition of interests between workers and managers, for whom the European continental description of Social Partners seems scarcely applicable. In the U.K. no-strike agreements are accepted only reluctantly, and one union in the U.K. watched the complete failure of a company and the loss of all its members' jobs, rather than conclude such an agreement. For both management and workforce, deep-seated fears of change and of their own ability to influence the consequences of change, are evident in much of the resistance to job design in more than one country.

Undoubtedly one of the reasons for the slow development of better working conditions has been the reluctance to use scientific knowledge. Senior managements are usually reluctant to consider areas of change in which they have no personal experience or are not familiar with in other companies. Day to day changes do not demonstrate any dramatic need for major alterations to management or authority structures, whilst long term trends are more readily put at the door of the market or some other external agency. It is a minority of managements who recognise the shifts in society and their implications for business and industry.

Unions are even more difficult to change—at least in Britain. The top officials have long exercised major power (political as well as industrial) and democratic changes from below are not always welcome. Recently some new General Secretaries and their supporters have modified their unions and the way they work—not always to the applause of their brother unions. Recently, too, some unions have recognised the benefits of employing competent researchers to assist in their discussions with management concerning working conditions, as well as to provide basic information for the formulation of policy. Even today, however, it is noticeable that British unions lack the scientific 'punch' and power to take the lead in proposing new forms of work organisation and are still reacting to proposals from management. This phenomenon has been noted by several researchers, as for example in a private report on Italian unions by IMITAX Ltd. (1978) and in papers given to an OECD seminar in Vienna in 1974 (OECD, 1975).

A result of this relative ignorance is that changes become the subject of oppositional debate rather than discussion—together with some otherwise unnecessary suspicion about motives.

This is a short catalogue of some of the matters which illustrate why work design is slow to change, and perhaps why the problems are often misunderstood. Terminology can take the place of understanding, for example in the widespread use of such terms as 'group working' or 'industrial democracy', as if they were solutions to the problems we face rather than just vague descriptive terms which mean different things to different people.

So, where do we go from here? Since we are talking about work design in practice, let us accept that there are many areas where knowledge is sparse and further research is needed.

One area in which we should all attempt to make a contribution is in the general need to change public opinions. It is still necessary to spread reliable information on *why* better jobs are necessary. Many do not believe that improved industrial or commercial performance will arise from better working conditions, not even that a designed integration of technical and human requirements is the route to optimum performance.

There is also a quite widespread belief that industry is for other people, and that it has little future role in the support of a modern advanced society. Reference to any statistics for national GDP and what contributes to it will demonstrate the falsity of that view, but the demise of the industrial contribution could well result from the restricting effect which this view has on the urgent need for job improvements. People will not wish to work at jobs which are unattractive and the availability of better jobs in non-industrial enterprises will deprive industry of the better educated and the more intelligent people, which is not good for the growth of high technology industry. The necessary integration between societal development and working life requires working life to change to match society. However, some recent directions in Britain are attempting to modify society to match concepts of working life which were current in the early part of this century—a process which is doomed to failure.

In order to pursue the process of beneficial change there are several steps which we might follow. None of these is new—they follow a well worn track of identifying and recognising the needs of people at work, particularly taking account of known problems and the evidence of success by others. The recognition must follow, in consequence, that improvements are possible, since a belief in the possibility of what we are trying to do is vital, and then we are in a position to create the means to achieve our purposes.

This all sounds very simple, although we know it is not so, but many companies *are* changing. Various steps may be suggested to help us on the way.

One method of demonstrating the feasibility and effectiveness of change is to take people to see good examples, and this has proved very convincing. To speak with others who have successfully introduced change is a surer way for people to recognise the realities than any lectures, videos or reading matter. These last provide good reinforcement, but to stand on a shop floor and say to a doubting colleague 'Don't argue with me, argue with him, because he's made these changes and his performance is better than yours' can be a powerful initial influence for opinion change. It gives the doubter a personal experience that work design changes are possible and effective.

Once the will to change is awake in management, and it must be clear and firm at very senior level, appropriate reasons for changing must be identified at all levels. The views at one level may be radically different from those at other levels, yet these must be brought out and discussed amongst all the parties involved. If these reasons are not agreed and clarified, as well as setting and accepting the criteria for change, people working for different goals can hinder the whole process.

The Working Party seems the best way to pursue this with small groups of people having the relevant skills and experiences. These should be given the task of finding solutions to the wide range of practical problems which will have been identified. A monitoring committee co-ordinates this activity, sets the timetable and is backed by senior management. This process will be slower than imposing a solution from a management analysis or consultant's report, but it has the advantage that it involves everyone who will, ultimately, have to make the changes work. Everyone then has influence on and some ownership of the conclusions and a major stake in achieving successful change.

During the preparation process outlined above, a major effect is the change in perspectives and relationships between the different groups involved. Underlying reasons for the changes must be clear if people are to cooperate to a successful conclusion; there can be no 'hidden agenda'. This implies a clear ethical stance on the use of people and the relationships between them in the company. Also the measures set up to identify whether company performance has improved must include ways of ensuring that people are not experiencing worse conditions as a result of the changes. For example, the introduction of many processes now creates the possibility of computer monitoring of people's activities. There are sometimes good reasons why this should be done, but it should never be covert. Agreement on methods and purposes is quite possible, and if management and employees cannot agree that they are satisfactory, then the change should probably not be made at all.

Recognition (from the Working Groups) of what needs changing, supported where necessary be expertise to suggest how things can be changed, should be accompanied by proposals for measuring the current and future states. The benefits of the changes will be recognisable by different groups in different forms. Output and quality are widely recognised, but there are other aspects of work which can be measured and which will indicate improvements to those more closely engaged in the company's clerical, service or manufacturing processes. These should be built into the change process at an early stage since attempts at retrospective measurement of many of these factors are usually unsatisfactory.

Although this summary may have little originality, the problems of work design changes are sufficiently difficult that a reiteration of these points cannot but be useful. It is expanded in the more detailed and extensive material in the following chapters and we hope that they will assist in the vital process of job and organisational change which is so desirable if work and society are to continue in harmony.

References

European Foundation (1989). *Four year rolling programme 1989-1992*. European Foundation for the Improvement of Living and Working Conditions, Dublin.

IMITAX Ltd (1978). Personal communication.

OECD (1975). Quality of Life at the Workplace. *Regional Trade Union Seminar, Vienna, Final Report*. OECD, Paris.

Chapter 2
But what are the issues in work redesign?

John R. Wilson

Department of Production Engineering and Production Management,
University of Nottingham, Nottingham, NG7 2RD

and Susan M. Grey

PA Consulting Group, Hyde Park House, 60a Knightsbridge, London, SW1X 7LE

Introduction

Even a cursory scan of the literature shows that the field we call work redesign comprises a vast array of theories, models, methods, criteria and recommendations. In part this reflects the fact that the area attracts the attention of workers from several different, if related, disciplines—occupational psychology, management science, ergonomics—each with differing perspectives, approaches and, perhaps, aims. Perhaps even more pertinent to explaining the disparate nature of the field, a basic tenet for many of us is that solutions are frequently situation-specific. Furthermore, there is generally great difficulty in providing the 'proof' of success or failure which would establish a certain approach as 'correct' or 'incorrect'.

The diversity of effort in work redesign is one reason why we are still asking a number of questions, which are relevant whatever the basic approach or domain of application. These include: Where do we draw the boundaries of what constitutes a work redesign initiative—at the task, job, role, environment, or organisation level? What are key factors at each level and how do they act? How do factors interact and are their effects complementary or conflicting? What are the intervening effects of individual differences and of organisational structures and production systems?

A number of issues will be raised in the course of this chapter and treated in varying depths. We look at the extent of coverage in work redesign—at the types of changes we are discussing. This is done through presentation of an outline model of the factors and outcomes which are of key importance. Then we go on to discuss the influences of what are often regarded as the more traditional ergonomics concerns of physical environment, work conditions, technology and the human–machine interface. Organisation structure and climate, and production control systems have a tremendous influence upon what is possible in work redesign and on how successful changes will be, and these are considered next. Amongst job characteristics, autonomy will be discussed, particularly in relation to impact of, and on, supervisors' attitudes. We also consider the position of the traditional concerns, which stemmed from a scientific management approach, of work standards, payment and rewards systems and so on. Finally the importance of the change process and the role of participation are emphasised.

During the discussion we draw upon various investigations in which we have been involved over the past few years. These have included medium-length investigations in libraries, retail outlets and a major communications company and, together with a colleague, Chris O'Brien, a long-term study at a car component manufacturing company. We work from the perspective of original backgrounds in occupational psychology and management, which lead us to take a wide, or macro-ergonomics approach to work redesign studies. Such an approach has led us to develop a framework of the factors and potential effects within work redesign (despite a certain caution about accepting the universality of any models), not merely carrying out exploratory, action-research development.

Coverage of work redesign

We talk within this paper of 'work redesign'. Although the terms 'job' and 'work' seem sometimes to be used interchangeably in the literature, we use 'work' to distinguish a macro-ergonomic approach, within which we embrace issues of interface, environment and support. In addition we include those issues typically central to job design (especially in 'enrichment' type activity)—job characteristics (or content) with work context factors as intervening variables. We use 'redesign' in preference to 'design' to emphasise the 'new' or 'change' aspect, and to distinguish the more modern, people-centred approaches from what was termed 'work design' within a scientific management perspective. Blackler and Shimmin (1984) seem at first to distinguish terms by talking of 'an impressive body of job-design theory, as well as a practical technology of work re-design...' (p. 105); however they seem subsequently to use job and work redesign (if not job design also) interchangeably. Wall and Martin (1987) similarly seem to draw little if any clear distinction between job and work design and redesign.

What then do we include under work re-design? The authors originally proposed a scheme of ergonomics-related factors and effects in work redesign at the Vancouver ODAM conference (Wilson and Grey, 1986). This 'model' originally also indicated routes for measurement and evaluation. Since then it has been simplified and used as the basis for explanations of how ergonomists use information from job holders and others in the design and evaluation of systems (Wilson, 1988). We have now tentatively suggested a further modification, this time accounting for organisation structure and prevailing culture, and (in manufacturing) for the philosophy or system of production control which is used. These factors, we are suggesting, act as both limiting and determining influences upon choices in job content, technology, interface etc., and also as intervening variables on the strength of the effects of the work re-design factors upon attitudes, performance and well-being.

Within the simplified version of the framework shown in Figure 2.1, certain outcomes in terms of better or worse performance, worker well-being and attitudes will be found as a result of:

- job and role content (characteristics)
- the technology (quality), human–machine interface and physical workplace, environment and conditions
- support procedures such as communications, training etc.
- work context (payment, job security etc.)
- interactions between these.

We are unsure as to whether the organisation and the production control system should be treated as factors at the level of the others, as 'parent' factors, or as intervening

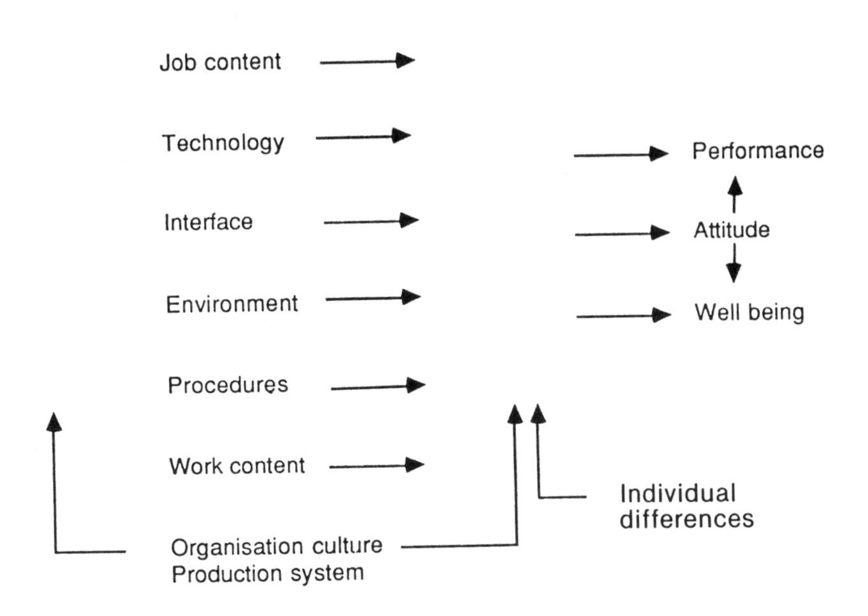

Figure 2.1. Framework for work redesign: factors and outcomes.

variables. Certainly individual differences act as intervening variables, with people's skill, knowledge and needs determining the direction and extent of any effects. Outcome variables or effects are grouped under those to do with performance (quantity and quality of work), worker well-being (satisfaction, health etc.), and attitudes. These latter are crucial. We have argued previously that, if workers perceive that they are provided with poorer than necessary working conditions, this can lead to attitudes that make successful introduction of any change impossible (Wilson and Grey, 1986). It can also have a profound impact on outcome variables related to performance and well-being. Although it may be simplistic, we have found that within this framework we can conceptualise, investigate and report the various issues of importance.

An interesting parallel to the development of the authors' thinking has been found amongst colleagues at the Social and Applied Psychology Unit, University of Sheffield. In their work, they classify human factors in advanced manufacturing technology into six areas of work: the systems design process, allocation of functions between people and machines, job design, organisation structure, and what they term hardware ergonomics and software ergonomics (e.g. Clegg and Corbett, 1987). A generous interpretation of our and their sets of areas or factors would see many overlaps and parallels.

We offer our outline framework with some caution. As proponents of an action-research approach to work design changes, and of the need for flexibility and learning on the part of investigators, we do have concern that over-emphasis on formal models and methods can produce a rigid, formal and autocratic approach, (see, for instance, Sell, 1986).

However, models and their accompanying instruments do allow comparative evaluations, better communication of results and recommendations, and give a certain scientific face validity.

Despite the doubts of many academic researchers, the most widely used job design model is the Job Characteristics Model and its accompanying Job Diagnostic Survey (Hackman and Oldham, 1980). These, like the earlier theories and concepts of job enrichment, seem to hold an attraction for industrial customers and practitioners. It may be argued that objections from ourselves and other academics—on the grounds of the job characteristics included, or the proposed relationships in the model, or assumptions of the invariance of respondent perceptions—are small beer compared to the fact that at least organisations are using a rational and repeatable approach to promote better job design.

Issues in work redesign

Influence of environment, technology and interface

Current theories of job design tend to concentrate upon characteristics of work intrinsic to jobs or roles. They allow for the moderating influence of organisational work context factors, but generally downplay the effects of the more tangible, physical elements in work. Technological variables have, however, been reported as having effects on various outcome variables (Brass, 1985; Pierce, 1984). We have previously noted that technical faults, inefficiencies and breakdowns have all resulted in hostile worker attitudes; these in fact found expression through dissatisfaction with easily grasped work context factors (Wilson and Grey, 1984). There is a strong link here, of course, to organisational policies and their effects on the relationship between work factors and outcome variables. Policies on equipment commissioning and the degree of involvement of workers, on preventive maintenance and repair, and on giving operators enlarged jobs requiring additional skills of fault diagnosis and rectification, will all moderate attitudes to technical systems and will influence the use of such technology and the performance obtained from it.

The interface—by which we mean both the human–machine physical interface comprising controls and displays and the cognitive interface comprising the information communicated both ways—we distinguish as a separate factor from the technology (although demonstrably in its physical sense it is a part of the technical system). Whereas the latter usually has influence by reason of any problems, smooth running passing often unnoticed, the interface design will affect the everyday work content and role of an operator. The degree of control and breadth of influence exerted by operators will be determined by the task behaviours (skill-, rule- or knowledge-based) the controls and displays will support (Rasmussen, 1986) by the extent to which the displays match, enhance and strengthen the operators' mental model (Wilson and Rutherford, 1987), and by the degree of display understanding and interpretation which we can promote. As a simple example, if we aim to give CNC/DNC cell operators the skills to be 'midwives' not 'monitors' (Wall, 1986), letting them perform some programming and fault rectification, then the display systems must support this. Thus there is a direct relationship between the human–machine interface and the outcome variables of performance and attitudes, and an indirect one in the way that the interface interacts with job content and characteristics.

Finally, in this set of physical influences, we have the physical conditions of work, environment and workspace. During most of this century psychologists and, later, ergonomists have argued the importance of physical environment effects on all three types of our outcome variables (see Schiro and Cox, 1978 or Sundstrom, 1986 for instance). This interest has not led to the general availability of validated scales dealing with workspace, access, seating, lighting, climate and noise within job design evaluation instruments. We would argue from the experience of at least some of our field investigations that conditions of work determined by the physical environment exert a tremendous influence upon worker attitudes towards management, co-operation with new work procedures, technology change, and many other aspects of work (Wilson and Grey, 1986). 'If they [management] give us a place like this to work in they obviously couldn't care less about us; why should I care to work properly or to help the new system?' was a typical comment during a study of data coders. The physical environment is frequently seen by job holders as intrinsic, not extrinsic, to jobs (Taber *et al.*, 1985), and as such, along with the technology and the interface, is an important determinant of satisfaction, internal and external motivation, as well as of performance.

Influence of organisation structure and policies

Job design and work design changes must be implemented as a part of, and within, an overall organisation strategic plan. However, many companies in the UK do not have any human resources strategy or function (e.g. Guest, 1987), and hence have no plan within which to incorporate job design developments. The dysfunctional effects of this are noted in the context of information technology by Blackler and Brown (1986) amongst others.

Our work in manufacturing industry has confirmed the importance of vertical integration within a company (removal of, or better communication across, hierarchical levels), in order to get congruence of goals when introducing group working or autonomous work systems. Such systems will affect the roles of everybody concerned whatever their position in the organisation. The degree of vertical integration it is possible to implement will be dependent upon the decision between systematised, discretionary, or developmental modes of organisation control (see for instance Slocum and Sims, 1980), and upon the mode and degree of communication employed.

A paradox that was found in the introduction of autonomous group working was related to organisational policies. Despite the underlying less autocratic philosophy of the new system of working, some more traditional, ostensibly outdated, facets of industrial organisation were found to be critical. For example, there was a considerable need felt for work performance standards, for status differentials to be considered, and for construction of an understandable payment system. Thus the paradox was in the need to balance group freedom and self-determination (in terms of goals, methods, work balancing and quality) with the prop of measurable concrete standards. Some degree of prescription was found to be necessary in order to oil the wheels of change at the start. Incidentally, the payment system proposed included a change in method of reward, not just in rates; new incentive schemes must pay for worker flexibility and acceptance of change. Dissension and doubt sown over payment policies perceived as confused or inequitable can sink a whole change initiative. Any attempts to change pay policies by sleight of hand or by obfuscation must be resisted. Clear, consistent and timely communication to the whole group affected, and often to all in the organisation is vital.

Perhaps the most extreme example of the need for traditional, prescriptive, perhaps more autocratic policies, came with the development of a Code of Practice. It was found that the (apparently) liberating approach of autonomous groups did in fact require a Code of Practice to be developed in order to allay suspicions, uncertainties and fears, at least at the outset. This Code covered personnel, maintenance and tooling, scheduling, quality, production engineering and training, but not payment which was covered elsewhere. The Code was developed collaboratively by various company personnel at different levels. The need for it was an example of a degree of authority being incorporated into a participative process. We can communicate and promote participation from the start, but cannot expect people suddenly to believe and act upon non-directive statements after years in a different system. There is a need for some prescription of the content of change (more so than for the change process—see later) at the outset. The Code of Practice also had benefits in providing clearer guidelines for the present and benchmarks for the future, and the process of constructing it 'forced' a re-evaluation of the change decisions made up to that time.

Influence of production systems

Job design changes in manufacturing industry will often be context specific. Generally, as Kolodny said at a 1984 SME Congress (quoted by Susman and Chase, 1986), job design will be production change driven and always production change oriented. Newer approaches to production, such as group technology or organisation for Just-in-Time, will provide both opportunities and also their own requirements for changes in work design.

Group technology for instance can provide a very convenient vehicle on which job redesign, in terms of autonomous groups, can hitch a ride. Increased autonomy can be a logical parallel with group technology; other generally desirable job characteristics can be enabled also—for instance task identity or skill variety.

The increase in application of Just-in-Time (JIT) principles in UK manufacturing industry has a number of implications for job design. In particular it poses a challenge to practitioners of job redesign in terms of the degree of its compatibility with the objectives of job enrichment and job enlargement. In many instances JIT serves to highlight the restrictive nature of the still pervasive practice of scientific management, which encourages high task specialisation, minimal training, individual financial incentives and directive supervisory styles. It is increasingly suggested that policies such as worker flexibility and participation, together with a higher level of shopfloor training, 'facilitative' supervision and a move away from high individual incentive payment systems are appropriate to the application of JIT on the shopfloor. In this sense, there is at least a superficial similarity between the objectives of job redesign and the personnel-related requirements of successful JIT implementation.

Autonomy and supervision

Enough has been written about job characteristics in the job design literature; we need not produce a full commentary here. We shall content ourselves with two observations. The first is to note the key role that 'communication' plays in the effects of many of the so-called key characteristics (see also Wilson, 1988). Feedback, from supervisors, peers, customers or the work itself is obvious but role clarity, task identity and significance, learning, and training quality will also all be determined in part by how good a company's communication policy and practice are.

The second observation pertains to autonomy—the key to many individual and group work design changes and important enough in our manufacturing industry case to warrant production of a code of practice to itself. This is intimately related to supervision and supervisory practices, especially in a (semi-)autonomous work group change. Cordery and Wall (1985) indeed believe that supervisory behaviour is the key route to implementing autonomous work practices. It is possible however that a broader brush approach is needed, with several mechanisms and practices set up to support autonomous working and its promotion. Autonomy is not a simple outcome; it does not just happen. Like a participative process it has to be directed, even if subtly. It also depends upon good in-company communication, and not on the typical approach of filtering and censoring information as it goes up the company hierarchy, and restricting but inflating it as it comes downwards. Autonomy also is not unitary; it has several different possible foci and each demands a different emphasis and different procedures. As examples, workers can have control over methods, targets and schedules, timing, criteria for acceptance of output and so on.

It should be added here that supervisory attitudes may often seem to be a block to work changes involving worker groups and autonomy (for obvious reasons of perceived status, fear, or uncertainty). However, doubts expressed by first-line supervision—although frequently seen as a bottle-neck—may in fact be well-founded. They are often the only people who know about the work, the technical systems and the capabilities and attitudes of the personnel involved.

Process of change

The over-riding task of the work redesign team in our manufacturing industry case was to ensure that both the process and content of change were geared to sustain the process itself. Mechanisms to do this were built from the outset, involving new company communications procedures, group and team meetings and political and structural flexibility to be able to 'change the change' if necessary. Only in this way, by setting up a robust, iterative and adaptive process, can we translate enthusiasm for work redesign in the abstract into practical developments that are contributed to and promoted by all parties involved.

Many things will aid in establishing a suitable change process: clear perceptions of the role of the change agent—and what happens when that agent leaves (a particular problem faced by one author); skills in project management; an acceptance that 'things change'; the establishment of shared goals and recognition of equity in the new system; recognition of the iterative nature of the process, and ability to give it time. These are all critical. Underlying much of the above is that the change, its mechanisms and development of its content, should be participative. However, it must be stressed that this is not as simple as slogans would have us believe. Quite apart from the fact that there are many different types and depths of participation, there are many potential pitfalls and problems. These may tarnish or even obliterate the oft-repeated advantages that participation leads to better solutions, greater commitment, and systemic spin-offs. Sometimes cynical use, even at times mere sloganising, has led to participation falling into some disrepute. Intrinsic problems of how to run or organise participation parallel the same issues for autonomy; participants must be slowly introduced to what is required, adequate resources of time and money must be given, and again good project management skills are required (see Wilson, 1990). Nonetheless, a philosophy and a practical set of policies and mechanisms for participation must underly most, if not all, successful work redesign change processes.

Conclusions

So then, what are the issues involved in work redesign? Our suggested framework of factors and outcomes has obviously widened the scope from the narrower base of issues traditionally taken as a part of job design theories and implementation. The influence of physical stress, for instance, on satisfaction and motivation (as well as on performance) in some circumstances will be greater than that of job characteristics. Certainly, we would argue that the many interactions apparent between factors dictate that an approach to work redesign as wide as the one we suggest must be taken. The role and job content of the operator cannot be viewed in isolation from the design of the interface. Organisation culture, technology choice, allocation of function, skill variety, autonomy and training procedures are also intimately linked.

Other issues in work redesign that are of importance, but which we have had no space to detail here, are international cultural influences, the effects of life out-of-work upon attitudes and behaviour at work, physical and mental workload, and the ethical and pragmatic considerations which underpin any decision as to how, and in what capacity, ergonomists should act in work redesign. Summarising the issues that have been raised and discussed in this chapter, and that have emanated from the practical work carried out by the authors, we find that:

- the redesign of people's work, and by implication investigations of attitudes to work, must include a wide range of factors of influence: job and role content, work context, support procedures, technical systems, interface, physical environment, production system and organisation policies and environment.
- individual differences must be allowed for, not just in terms of the intervening variables of skills, knowledge and needs, but also in terms of the different perspectives that will be found between, say, line and service management, first line supervision and shopfloor workers, skilled and unskilled, old and young.
- change must be relatively slow, iterative and participative. However, it does require certain direction and prescription, especially at the outset when dealing with a naturally sceptical, suspicious, or fearful workforce.
- change must also be flexible; things change often in a dynamic organisation and so must any work redesign strategies. Implementation is a learning process for company, management and workforce, and so it should be for change agents also.
- successful work redesign is very dependent upon the skills and personality of the change agents. Although they must have appropriate inherent personal characteristics, they also need training and experience in project management. We must also avoid the 'Tito syndrome', by leaving in place the means to continue the change process after the initiating agent has gone.
- most importantly, the vital issues in work redesign are not so much the content and boundaries of the change, but where we start, how we proceed and how we sustain the change. The over-riding role of the change process is to sustain the process itself.

References

Blackler, F. and Brown, C. (1986). Alternative models to guide the design and introduction of new information technologies into work organisations. *Journal of Occupational Psychology*, **59**, 287–313.

Blackler, F. and Shimmin, S. (1984). *Applying Psychology in Organizations*. (London: Methuen.)

Brass, D.J. (1985). Technology and the structuring of jobs: employee satisfaction, performance, and influence. *Organisational Behaviour and Human Decision Processes*, **35**, 216–240.

Clegg, C.W. and Corbett, J.M. (1987). Research and development into 'humanizing' advanced manufacturing technology. In: *The Human Side of Advanced Manufacturing Technology*, edited by T.D. Wall, C.W. Clegg and N.J. Kemp. (Chichester: John Wiley & Sons).

Cordery, J.L. and Wall, T.D. (1985). Work design and supervisory practice: a model. *Human Relations*, **38**, 425–441.

Guest, D.E. (1987). Human resource management and industrial relations. *Journal of Management Studies*, **24**, 503–521.

Hackman, J.R. and Oldham, G.R. (1980). *Work Redesign*. (Reading, Massachusetts: Addison-Wesley.)

Pierce, J.L. (1984). Job design and technology: a sociotechnical systems perspective. *Journal of Occupational Behaviour*, **5**, 147–154.

Rasmussen, J. (1986). *Information Processing and Human-Machine Interaction*. (Amsterdam: North-Holland.)

Schiro, S.G. and Cox, G.H. (1978). A new approach to the evaluation of ergonomics changes. *Proceedings of the 22nd Human Factors Society Annual Meeting*, 479–484.

Sell, R. (1986). Attitude surveys: an authoritarian technique masquerading as a participative process. In: *Contemporary Ergonomics 1986*, edited by D.J. Oborne. London: Taylor & Francis. pp. 44–48.

Slocum, J.W. and Sims, H.P. (1980). A typology for integrating technology, organization and job design. *Human Relations*, **33**, 193–212.

Sundstrom, E. (1986). *Work Places*. (Cambridge, UK: Cambridge University Press.)

Susman, G.I. and Chase, R.B. (1986). A sociotechnical analysis of the integrated factory. *Journal of Applied Behavioural Science*, **22**, 257–270.

Taber, T.D., Beehr, T. and Walsh, J. (1985). Relationships between job evaluation ratings and self-ratings of job characteristics. *Organisational Behaviour & Human Decision Processes*, **35**, 27–45.

Wall, T.D. (1986). *Advanced Manufacturing and the Design of Operator Jobs*. Memo No. 809, MRC/ESRC Social and Applied Psychology Unit, University of Sheffield.

Wall, T.D. and Martin, R. (1987). Job and work design. Chapter 3 of *International Review of Industrial and Organizational Psychology* edited by C.L. Cooper and I.T. Robertson. (Chichester: J. Wiley & Sons) pp. 61–91.

Wilson, J.R. (1988). Communications issues in occupational ergonomics. Keynote paper in *Proceedings of International Conference on Ergonomics, Occupational Safety and Health and the Environment*, Beijing, October, pp 13–19.

Wilson, J.R. (1990). Participation—a framework and a foundation for ergonomics. *Ergonomics*, **33**, (submitted).

Wilson, J.R. and Grey, S.M. (1984). Reach requirements and job attitudes at laser-scanner checkout systems. *Ergonomics*, **27**, 1247–1266.

Wilson, J.R. and Grey, S.M. (1986). Perceived characteristics of the work environment. In: *Human Factors in Organisational Design and Management-II*, edited by O. Brown and H. Hendrick (Amsterdam: Elsevier).

Wilson, J.R. and Rutherford, A. (1987). Human interfaces with advanced manufacturing processes. Chapter 4 of *International Review of Industrial and Organizational Psychology*, edited by C.L. Cooper and I.T. Robertson. (Chichester: J. Wiley & Sons,) pp. 93–115.

Part II

Ergonomics practice in industry

The three papers in this section deal with different approaches to ergonomics studies in the industrial context. They identify various practical problems which may be encountered from the organisational, methodological, and implementation points of view. One issue which is clear from all three papers is that ergonomics techniques are still lacking for identifying changes which need to be made in the workplace, particularly as most industrial problems are multifactorial in nature. This can only be tackled by further research into the interacting influences of stressors at work, but will also need developments in methodology to take account of the less clearly defined factors acting outside the laboratory situation.

Ergonomic practice involves many others besides the ergonomics practitioner. It is necessary both to communicate and to work with engineers, designers, managers and equally importantly with the operators themselves. The difficulty of relating ergonomics knowledge to real situations is compounded by the need to translate this into the technical and economic terms used in the organisation's management and control systems. Change is expensive and must be justified. It is important to demonstrate the effectiveness of the ergonomic approach to those who have the power to introduce the changes, but evidence in the literature is scarce. Too few of the published case studies have taken the time to evaluate their success in terms of costs and benefits, and in addition it is not clear that we have yet developed the proper measures for evaluating ergonomic changes. As Aarås has shown in his paper, musculo-skeletal disorders are still difficult to control, even in ergonomically designed workplaces, and much better measures are needed to study psychological and sociological influences on work design.

The papers by Eklund and Haslegrave also contrast ergonomic practice in work design and work redesign. The basic principles may be the same, but practical problems differ and the constraints on solutions can be much greater in changing existing work situations. However, incorporating ergonomics in the early stages of the planning process can allow the design of buildings and equipment to be properly matched to the work organisation. In both situations, the extent of the ergonomist's influence is determined by his/her position as a consultant or as a member of a company team. Thus the organisational structure of the investigation can be an important factor in the success of interventions.

Chapter 3
Experience from ergonomics planning of industrial plants

Jorgen A.E. Eklund

Department of Industrial Ergonomics, University of Technology, S-581 83 Linköping, Sweden

Introduction

Very few investigations and research reports can be found in the literature concerning methods and strategies for incorporating ergonomics in the planning of new industrial production plants. Work regarding the design and planning process for buildings and user participation has been performed by Nilsson-Etzler (1978) and Steen and Ullmark (1982) among others. Very little can be found about the systematic ergonomics planning of production systems (Mårtensson and Veibäck, 1980), including the choice of production technology and work organisation, although references concerned with environmental factors can be found.

In this chapter, experience from several projects and production plants is discussed. The author has adopted the viewpoint of an applied ergonomist, and has also interviewed participants in different projects. The chapter is based mainly on case studies related to the planning of new engineering companies employing between fifty and sixteen hundred workers. The aim is to convey experience that other ergonomists may apply in similar situations.

The initial stages of a project

The position of the ergonomist is important for the possibility of influencing the results of the project. The ergonomist may be employed by the company, assigned as a consultant, or may act as an adviser from another organisation. It is clear that the person employed by the company is normally given responsibility for the results and therefore also has more authority to influence them. The consultant is often given the task of solving problems, which may include working methods defined by a non-ergonomist in the company. The adviser normally has few opportunities to obtain sufficient time and access to relevant information, and has less influence on decisions since he or she is often not present when the decisions are taken.

The possibilities for influencing the results are best in the early stages of the project, and will thereafter decrease continuously. It is therefore of great importance that the ergonomist should be able to participate from the start of the project in the formulation of policy for work conditions. It is of the utmost importance to come to an agreement about a written policy to avoid inferior quality solutions when unforeseen difficulties and costs are later encountered. The limits and ambition level are set by company management, and will of course be influenced by company operating conditions. Leadership attitudes, company culture and image play important roles in this respect.

In the earlier stages of the planning process, the basic philosophy of the new or redesigned plant is decided upon. For larger companies this often includes the basis of production system, design and layout of buildings, work organisation and personnel policy. Some companies have also included a social specification concerning desired attributes of the work tasks and workplaces. Unfortunately, owing to the traditional view of projects, buildings, production layout and technology are planned first, and work organisation and personnel policy later. Experience shows that the order should be reversed, because many of the aims from work organisation and personnel policy necessitate specific solutions for buildings, production technology and layout. By tradition, the industrialist views ergonomists as experts to be engaged in the last stage of the project, making final adjustments to the system if necessary. However, in some companies the benefits of including ergonomists throughout the whole project have been recognized.

The planning process

In small companies, planning is often the responsibility of the managers and engineers in the company, and few or no consultants are engaged. The planners often lack experience from other projects, seldom consult an ergonomist and are therefore less likely to integrate work environment aspects. Work methods are based more on their own experience and intuition. Larger companies produce detailed specifications for buildings, computer systems, machines, equipment and accident prevention. However, it should also be mentioned that it is not unusual to let a specialist company take the total responsibility for planning, purchase and installation, in other words delivering a 'turnkey' sub-system, production system or even production plant, which means that there is less demand for detailed specifications. Psychological and social specifications for jobs and workplaces are becoming increasingly common in order to fulfil the needs of the workers. These specifications are of great importance, but also very difficult to produce. There is little or no material in the literature about any type of specification. There are also extremely few evaluations or summaries of experience from earlier or specific projects.

There is documented experience that user participation is beneficial. Experts do not have as much knowledge about the specific requirements of the jobs as the workers themselves. It is essential to test equipment, prototypes and workplace mock-ups with the users in realistic situations. The quality of the results has been vastly improved in many cases and attitudes towards the solutions and new work situations have become much more positive as a result of user participation. User participation is also an important part of the implementation strategy, but this has not yet been accepted and put into practice by most companies.

Many individuals influence the planning process. A very important factor in the Swedish context is the participation of unions and union representatives. Government and local authorities influence the introduction of specific solutions. Legal requirements may sometimes support good ergonomics solutions, but may also create criterion limits and indirectly support the choice of minimum levels.

Within a company there are sub-groups representing different specialist functions, sometimes including ergonomics or occupational health. These separate departments have different goals and viewpoints. The proposed solutions therefore often create a conflict of interests, and it may be difficult to reach good ergonomic solutions. This is another reason why ergonomics must be a management strategy, defined from the start of a project, instead of being negotiated through a complex and restricted planning process. One major task for the ergonomist is to formulate proposals or specifications for proposals and then to 'sell' the ideas. For this purpose, many examples from other companies are needed as arguments, together with calculations of effects and evaluations of consequences.

Discussion and conclusions

One of the conclusions that can be drawn is that ergonomists and practitioners are in need of case studies and documentation from many projects and production systems. This includes methods and strategies for managing projects, as well as for involving ergonomics on the same basis as production engineering, quality and maintenance. There is also a need to develop the use and spread of ergonomics requirements and specifications. In particular methods of evaluating psychological and sociological consequences are badly needed for alternative production systems already in the planning stage. Finally, evaluations of planning projects and their results are needed in order to obtain better knowledge about these processes.

Participation in industrial planning imposes very different demands on the ergonomist from those in research, development and other industrial assignments. Little support can be obtained from the literature. Present ergonomics education hardly deals with these matters and there is also a severe shortage of ergonomists with extensive experience in the field. If ergonomists in the future are to be competitive in this area, much more emphasis on evaluation and documentation is needed, as well as conferences, seminars and education.

References

Mårtensson, L. and Veibäck, T. (1980). *Nya stålverket i Hofors. Planeringen av arbetsmiljö (The new steel plant in Hofors. The planning of the work environment)*. Rapport TRITA-AML-0097, (Stockholm: Arbetsmiljölaboratoriet, Tekniska högskolan), (in Swedish).

Nilsson-Etzler, B. (1978). *Tidig projektering av industriell arbetsmiljö (Early planning of the industrial work environment)*. Rapport R 37:1978, (Stockholm: Statens råd för byggforskning), (in Swedish).

Steen, J. and Ullmark, P. (1982). *De anställdas mejeri (The Employees' Dairy)*, (Stockholm: Libertryck), (in Swedish).

Chapter 4
How well can ergonomists address problems identified in the workplace?

Christine M. Haslegrave

Department of Production Engineering and Production Management,
University of Nottingham,
Nottingham, NG7 2RD, U.K.

Introduction

Consultant ergonomists are often called in to advise industrial management on the redesign of workplaces, when they are essentially acting in the role of troubleshooters. In this situation, they do not have the advantage of being 'in' at the start of a project and, because of this, selling worker-centred design is probably even harder than at the production planning phase. They are usually asked to look at a specific problem and to give an instant solution. The problem appears already to have been identified and the company are expecting a simple response to rectify this.

The actual situation is rather different. The management's problem is certainly real, but the ergonomist knows that the true causes (and therefore solutions) are likely to be more complex than anticipated. However, if the cause is not immediately apparent, it is not feasible to embark on a large scale investigation to find a complete answer to the problems of one small company. Moreover, any investigation on the shop floor has to cause the minimum disruption to production and thus cannot usually involve controlled experimentation.

This creates a series of problems for the consultant ergonomist, which are not only methodological but are also related to the theoretical framework and knowledge base of ergonomics. Some of the problems encountered in practising work redesign can be illustrated by a small case study which was carried out for a production sewing company. It also provides an example for discussion of the techniques which are currently available for field investigations, and of how successful ergonomists can be at present in making changes to the design of workplaces and of work.

Case study from the production sewing industry

The study was typical of many which face ergonomists when they are called in to give advice. It was initiated by the Production Manager because a few of the sewing machinists (all women) were experiencing wrist problems. Such problems are very common in the sewing industry, as has been well documented by Punnett *et al.* (1985), and in several other industries as well. From the manager's point of view it was a fairly limited brief—he was concerned about his employees and from his limited knowledge of ergonomics he saw it simply as a question of improving the layout of the machines.

In order to investigate strain injuries of this type, it is necessary to study the task in some detail. However, the company had several production lines made up of perhaps fifteen to twenty workstations, each line making a slightly different product. A typical workstation is shown in Figure 4.1. The task factors and exposure of the groups of machinists could vary significantly between the different lines, so that an overall analysis including the whole workforce was not possible.

Figure 4.1. Typical sewing workstation

Approach to the investigation

As with many such studies, resources were limited and it had to be a very quick two to three day exercise. The investigation involved a relatively standard battery of tests:

Checklist of workplace layout and work organisation factors
Epidemiology
Questionnaire survey and interviews
Observation
Video record of tasks
Environmental measures—lighting, noise levels

A proper epidemiological analysis was not possible with the small numbers of operators who were affected, so that inferences about possible causes of the injuries had to be drawn initially from the few medical reports. In such situations, both proper analysis of the causes and assessment of the benefits of workplace changes are extremely difficult (and it can also be difficult to convince management of the validity of findings).

Since the injury data could not be used as reliable indicators of the causes of the problems, other methods of investigation had to be found. These were based on the assumptions that discomfort or pain is a precursor of injury and that information on the lower levels of discomfort could be used as a good indicator of problem areas. All operators on the affected production line were therefore asked to complete questionnaires about their work and a sample were later interviewed.

Questionnaires and interviews

The questionnaires and interviews gave a great deal of information about the work the machinists were doing, about their experience of fatigue and discomfort over the working day, and on how the layout of the workplace influenced the way they could perform the task. This was very useful in realising the extent to which the design limitations of the workplace and equipment hampered them in their work. The video records clarified this at the more detailed level of the individual hand and finger movements. Even more importantly, the information given by the operators very quickly convinced the ergonomists that the problems identified by the management were only a very small part of a much larger problem.

Two examples of the responses to the questionnaires, shown in Figure 4.2, can be used to illustrate some of the aspects of the task. This section of the questionnaire was concerned with experience of any discomfort or pain (below the level which would cause reports to medical staff). The first question was:

'In the last 4 weeks, have your noticed any feelings of pain, ache, discomfort or excessive tiredness in any part of your body?'

If so, the operator was asked to mark the areas on the body diagram (adapting the technique developed by Corlett and Bishop, 1976). Having done so, she was asked how frequently the pain or discomfort occurred, when it occurred during the working day and working week, and whether it was associated with any particular activity.

From this information it was possible to obtain a rating of severity of the aches and pains for different jobs within the factory, using measures of the frequency of discomfort and time of onset during the day. This method has proved quite sensitive in indicating which jobs present the greatest problems for operators and also in identifying potential causes within the workplace.

The responses to the body diagrams in the questionnaires showed that the machinists were able to identify very specific sites of discomfort. The first respondent in Figure 4.2 reported an unusual number of aches and pains. However, other machinists reported very similar problems and, when taken as a whole picture, their symptoms could be related to different aspects of the task. Reports of back pain and leg aches are commonly symptomatic of poor working postures, in this case caused to a large extent by poor seating. Elbow pain and shoulder fatigue in the left arm were associated with moving material across the sewing table.

The machinists used their right arms to feed the material through the sewing machine. Several grips were used during the sewing tasks (for holding and pleating material, and for tying off or sealing threads). These grips were very similar to the ones reported by

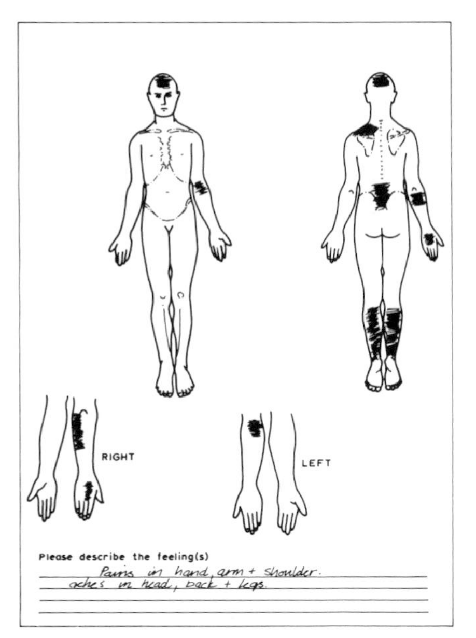

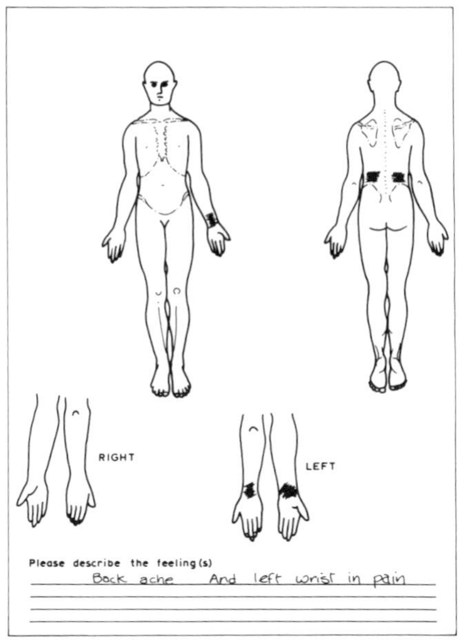

Figure 4.2. Questionnaire responses related to pain or discomfort experienced at work

Punnett and Keyserling (1987) in the garment industry in the United States. As demonstrated in Figure 4.2 the first machinist had muscular pain in a particular region of the back of her fingers, which she experienced every day in the afternoon. Other workers had pains in various other parts of their hands, not all in the same hand. In the study, there were in fact many more complaints of finger pain than of the wrist problems which the management had identified.

The arm and hand motions were complex and combined extreme ranges of joint movement with moving or lifting material in a highly repetitive set of tasks. These have all been identified as risk factors for strain injuries.

As a final point on the questionnaire responses in Figure 4.2, the area marked on the machinist's head was intended to represent headaches and tiredness. This illustrates the complexity of the issues in work design. It is not simply concerned with single task factors. Sewing production lines are noisy and require high local lighting which may be glaring. The enclosed work stations were hot and the continuous movement of material meant that the air was dusty, so that environmental factors were likely to cause fatigue and stress, and also to influence performance and motivation.

Outcome

The women were working on a partially automated production line in which the supply of material was mechanised, so that they very rarely had to leave their seats, except perhaps every two hours in order to replace the reels of thread. The workstations were separated and enclosed by buffer stocks of material. This effectively prevented gossip or indeed any work-related communication without stopping work. A secondary effect was that ventilation was reduced, making the workstations much hotter than they would have been in a more open layout.

The obvious ergonomic problems identified in the workplace are listed in Table 4.1.

Since there had been recent changes to the level of automation on the production line, the assessment concluded that further monitoring of the situation might be necessary as the effects of the changes became more apparent, particularly as the pace speeded up when the operators became familiar with the new system.

Table 4.1. Ergonomic problems identified

- layout of sewing table, pedals and sewing head
- poor seating with insufficient adjustment
- interfacing of automated material delivery and workstation
- lack of variation in posture and task
- local illumination of task
- physical environment
- social environment with isolated workstations

Causes of Injuries

The injuries which had been reported seemed to lie in the motions of the hand and particularly in the wrist and fingers. Reference to the literature on the current knowledge of strain injury problems produced a range of possible recommendations for improvements to the work design (listed in Table 4.2), but none of these are specific to any task. In a particular work situation it is difficult to judge which is the most important, or whether all are necessary or only one.

In the present study, the hand motions involved relatively low forces, but were highly repetitive, the hand postures being very cramped. There were several levels of repetition in the job, since the operation of sewing a single seam itself required highly repetitive individual movements of the hand and arm. There was no way of knowing which elements of the task were causing the problems. A solution to the problem cannot be found simply by changing the seat and sewing table, although this would of course improve the working posture. It is necessary to make more radical changes in the working method.

Table 4.2. Work redesign recommendations to reduce strain injuries

- reduce extreme joint postures, pace of work and/or forces
- increase variation in work
- introduce frequent breaks away from the workstation
- introduce exercise breaks to relax muscles
- improve social interaction

For an immediate solution, the only recommendations which could be made to the management were to reduce the pace of the work or to increase the variation in tasks. Without further research, which would take a long time and could not be supported by an individual company, it was not possible to propose changes in the way the material was held (detail of working methods) nor to be sure how proposals to redesign the sewing machines might affect these.

It was possible (although unlikely) that the fact of change itself could have caused the wrist problems, through a temporary increase in stress, (e.g. fear of redundancy or difficulty in coping with new procedures) or even because the machinists had to retrain to some extent (learning slightly different hand movements). This type of question can only be resolved by long-term follow-up studies. The solution to the injuries experienced by the machinists was obviously not simply that seen by management of quick modifications to the workstations. That was the first step, and important in itself, but it could not solve all the problems of the operators immediately.

Inadequacy of measurement techniques

In addition to the difficulties in prescribing modifications to improve the workplaces, the study also raised some questions about the techniques ergonomists have available to investigate problems in the field.

There were methodological problems in defining and recording hand postures, and in estimating the duration of potentially harmful postures during each work cycle. Hand movements in this type of work are more complex than those which have been studied in the laboratory, and are difficult to analyse in terms of individual joint motions. They are also highly three-dimensional and thus difficult to measure from video film, or even to film on the shop floor.

Highly sophisticated laboratory-based equipment cannot easily be used in the field. If it is introduced into the industrial workplace, it is likely to hinder or change the working methods. As far as physical aspects of the workplace are concerned, the situation may be improving with miniature transducers and automated and accurate analysis of video records, but instrumentation for evaluation of the mental components of work in the field is limited. More seriously, there is still a large element of experienced judgement and intuition in identifying problems in the workplace. The use of laboratory-based equipment may give more accurate recording, but it is often likely to be too specific in its applications to address questions such as the identification of the true source of a problem in a complex working environment. However, subjective measures (such as the questionnaire techniques used in this study) can be very valuable for investigating multifactorial work problems although they are sometimes undervalued, especially by industrial clients.

Interactions between factors

The present study was concerned with a problem (strain injuries) which, despite intensive investigation, it not well understood. Nevertheless, many other human processes and malfunctions are also not yet properly investigated, and the ergonomist very rarely has an immediate solution to a real-life problem.

There was a second problem in this study in that it was a dynamic situation in which the company was introducing 'improvements' as they automated different parts of the production, but the ultimate effects of the changes could not be predicted. It was possible to hazard a guess that some of the changes were only likely to make the problems worse, but the extent was quite unknown.

There were obviously many physical, environmental and psychological factors involved, but ergonomists do not yet understand their interactions, particularly in relation to the social and psychological environments of work. Of course, the problem of strain injuries is notorious for its multifactorial nature, but it is again not unique in this. Many other jobs of a less repetitive nature are equally affected by a multiplicity of stress factors. The basis of ergonomics is experimental trials, but it is possible that these have been confined too closely to studying one or two variables, assuming that the relationships hold generally. Changes in a workplace rarely affect only a single factor, and so can sometimes be difficult to relate to textbook data.

In many industrial studies there is also a also a danger of taking the original problem posed by the management at its face value. What may seem a reasonable question from their perspective, can in a systems approach be seen from quite a different angle. In the present case, the management reported wrist problems and naturally associated these with the primary right-handed task of feeding the material through the machine. The full ergonomic investigation found more finger problems than wrist problems, and the wrist problems themselves were more generally related to lifting material with the left hand.

There is a particular difficulty when young newly graduated ergonomists work in industry, whether as employees or as consultants. They are liable to take a narrow approach to ergonomic evaluations and to consider only single factors of a complex problem. Incomplete or unsuccessful work modifications can damage the ergonomics profession as a whole, because the managements very quickly realise that a study is not effective, although they may not understand the true reasons for this.

More attention needs to be paid to this aspect in the training of ergonomists. It is not enough to give them the theoretical knowledge. The need for practical experience is probably more acute than in most other professions (except perhaps for medicine where there are many similarities). This is because ergonomics can never be an exact science. The complexity of the interactions involved, and the detailed aspects which are so task specific, mean that it is essential to interpret the necessarily limited information which can be obtained from an existing workplace and to qualify recommendations in the light of variability and imprecision.

Ergonomists have to learn to deal with variability in a way that most engineers never do, and it is necessary to develop expertise in these skills, as well as being able to justify and defend the underlying assumptions to engineers. Ergonomists are not well trained for this at present, and it may be important to develop ways of giving them accelerated experience as part of ergonomics education.

Conclusions

Three questions concerning the techniques which are available for evaluating work situations arise from the considerations illustrated by this case study:

How do we identify the real problem in the workplace?
How do we deal with work techniques which are more complex than the ones that have been studied in the laboratory?
How can we be sure that we are dealing with physical problems, and isolate these from the mental effects?

The analysis suggests that new instruments and measures are still needed to answer these. On the complementary aspect of the use of existing knowledge, there is a massive body of literature on ergonomics studies and databases. Many of the solutions to workplace problems should be immediately available, yet (to take the case outlined here) ergonomists can approach a relatively simple problem in industry and still be able to make only partial improvements.

Ergonomists involved in work (re)design urgently need to consider some of the practical aspects of implementation, and where the major effort should be directed to develop the discipline. There are several alternatives: to enlarge the databases, to study specific tasks (assuming that each problem is different), to improve techniques for field measurements, and to develop a theoretical understanding of the nature of industrial work which can predict the complex interactions between the physical, mental and social factors.

It is, of course, possible at present to make substantial improvements to industrial working conditions from current ergonomic knowledge but, without developments in both theory and techniques, further progress in improving working conditions will be difficult. Advances in any of the four directions will require considerable effort and resources. The third alternative of improving instrumentation is of course a continuing process, while the fourth is a long-term goal which will inevitably take a considerable time as in any other developing science. The dilemma at present for the practical application of work design is the very real choice between the first two alternatives in deciding where to invest scarce research funds to develop the knowledge base which is so crucial to ergonomics practitioners.

References

Corlett, E.N. and Bishop, R.P. (1976). A technique for assessing postural discomfort. *Ergonomics*, **19**, 2, 175–182.

Punnett, L., Robins, J.M., Wegman, D.H. and Keyserling, W.M. (1985). Soft tissue disorders of the upper limb in women garment workers. *Scandinavian Journal of Work Environment and Health*, **11**, 417–425.

Punnett, L. and Keyserling, W.M. (1987). Application of critique of job-site work analysis methods. *Ergonomics*, **30**, 7, 1099–1116.

Chapter 5
Load-related musculo-skeletal illness: is ergonomic workplace design a sufficient remedy?

A. Aarås
Standard Telefon og Kabelfabrik A/S, Østre Aker vei 33, 0581 Oslo 5, Norway

Introduction

An epidemiological study was carried out on thirty-seven female workers who performed electro-mechanical assembly tasks at work places designed according to ergonomics principles, as described in Aarås *et al.* (1989). Postural load was measured together with the duration of low activity levels. The median value of postural load (in terms of static trapezius load measured by electromyography (EMG)) was low for the group of workers and did not exceed 1% to 2% MVC (Maximum Voluntary Contraction) for most of the working day (Aarås and Westgaard, 1987). The static trapezius load was quantified according to the procedure described by Ericson and Hagberg (1978), which defines the muscle load as the level exceeded for 90% of the recording period.

A standing posture was adopted by the workers for most of the working day. For this posture, the group of workers had a trapezius load below 1% MVC for less than 5% of the total recording time and below 2% MVC for less than 12% of the recording time calculated as a group median value (Aarås *et al.*, 1989). The trapezius load was below 5% MVC for about 50% of the recording time (Aarås, 1989). These were calculated by counting the total duration of crossings below 1, 2 and 5% MVC from the digital full wave rectified and integrated trapezius signal.

Shoulder loading was assessed by measuring the median flexion/extension of the upper arm in the gleno-humeral joint (i.e. the value of the angle for 50% of the recording time). This was usually less than 15°, while the abduction/adduction was mostly less than 10° (Aarås *et al.*, 1988). The duration of low shoulder moment, defined as flexion/extension in the gleno-humeral joint between +5° and −5° (as a median group value), was mostly less than 20% of the time (Aarås *et al.*, 1989). Real breaks in the load pattern in terms of spontaneous pauses varied for individual workers and amounted to 30 to 60 minutes for each worker in an eight hour working day (Aarås, 1987).

These studies, therefore, showed the levels of postural load in the assembly tasks. However, other factors also influence the load on the musculo-skeletal system and the incidence of musculo-skeletal illness (Kvarnstrøm, 1983), including work load and stress factors outside plantwork, and these influences were investigated further.

The consequences of musculo-skeletal illness, in terms of sick-leave, were in fact low for these female workers, (Aarås and Westgaard, 1987). The level of musculo-skeletal sick-leave due to shoulder injuries was very similar to the incidence of such illness, regardless of body location, in a control group of female workers who did not have continuous work load (Westgaard and Aarås, 1984).

However, it is also important to consider other health parameters related to musculo-skeletal complaints, even when they do not lead to a period of sick-leave, when suggesting an acceptable work load on the musculo-skeletal system. The study investigated both intensity and duration of pain during work, assessing these by interviews. Signs of musculo-skeletal disorders noted during clinical examination of the workers gave further information on the development of such problems.

All these factors related to work load and the consequences for health must be assessed in order to evaluate whether the ergonomic design of work places is a sufficient remedy in terms of the incidence of load-related musculo-skeletal illness.

Material and methods

The subjects in the study were thirty-seven full-time female workers who assembled telephone exchange racks at three different systems 8B (8 subjects), 10C (11 subjects) and 11B (5 subjects). In addition, there was a mixed group of thirteen workers mainly from the 10C and 11B systems. The mean age of the subjects was 35.5 years, with a range of 24 to 52 years at the time of the interview in 1984.

Their work load outside the plant was estimated by asking the subjects about work and work load at home. Questions were asked concerning the division of work between the spouses, the number of children needing care, spare-time activities and physical exercise. Questions were also included concerning psychological problems, tension and sleep problems. A three-point ordinal scale was used to quantify all factors in terms of 'little' (none), 'some' and 'much' influence of these factors. The financial and housing situation for the subjects was also assessed as 'good', 'satisfactory' or 'bad'.

The reliability of the scales was established by interviewing nineteen randomly chosen subjects twice, about three months apart. The reproducibility of the answers to the questions above was in most cases 5% or better, in terms of difference in their responses.

Pain intensities were explored in relation to work capability, and assessed for the previous twelve months, the whole period of employment and the period before recruitment. Pain intensity was classified according to Westgaard *et al.* (1982).

The reliability of the pain scale was also tested by interviewing nineteen subjects about three months apart. This indicated a 2–3% difference in responses of one scale point and 1% difference of two points between the first and the second interview. It is also interesting to note that these results corresponded fairly well with the reported frequency of 5% differences in responses for the Visual Analog Scale (VAS) (Larsen *et al.* (1985) and Jensen *et al.* (1986)).

The duration of painful symptoms was quantified for the last year before the interviews in 1984, and recorded in accordance with the Standardised Nordic questionnaires for the analysis of musculo-skeletal symptoms (Kuorinka *et al.*, 1987).

Clinical examinations were carried out directly after the personal interviews. These consisted of a general observation of the musculo-skeletal system, measurements of the range of passive movements of the neck and head, palpation of muscle spasm and sore spots (trigger points) of trapezius muscle (Sheon *et al.*, 1982). Palpation of tendon attachments to supraspinatus and deltoideus was performed with relaxed muscles and against active resistance during the muscle contraction. Tenderness or pain when palpating the tendon attachment was recorded as positive.

Isometric and endurance tests were performed by:

1 Lifting the shoulders with the upper arms hanging relaxed beside the body.
2 Abduction of the upper arms to 90° and then further abduction.

The isometric tests involved maximum contraction against resistance and prolonged contraction against resistance for 15 seconds. The endurance tests were carried out by holding a 'smaller' muscle contraction against the weight of the limb and body part, i.e. low force of long duration. Thus, the endurance test attempted to simulate the physiological work pattern. If the subject felt tender or experienced continuous pain for more than a minute after cessation of the test, the result was classified as positive.

The examination of the spine was performed according to a screening procedure used by Department of Orthopaedics at Sahlgren Hospital in Gothenburg.

All these factors were statistically tested using the Wilcoxon test for tied observations, and the different work groups were compared for duration of pain in the year previous to the interview, clinical symptoms and signs, and psychosocial factors.

Results

Work load and important stress factors outside plantwork

The interviews assessing work load outside the plant showed that most married workers got good support from their spouses in doing the necessary housework. However, between 14 and 23% of workers within the groups reported that they had much work to do at home, as shown in Figure 5.1. In addition to housework, some workers ran farms and took care of one or two children. The rest of the workers reported having some housework and very few had little work outside the plant.

Psychological problems, also shown in Figure 5.1, may influence the development of musculo-skeletal illness and were reported by 30 to 46% within the groups. Every fifth or sixth worker suffered serious psychological problems which had an adverse influence on their mental well-being.

Good family situations were reported by more than 80% of the workers. Three workers out of thirty-seven described relations with the rest of their family as only satisfactory, and three workers had a 'bad' relationship with their family.

Sleep problems were rare (Figure 5.2), but 20 to 25% of workers reported occasional sleep disturbances. The feeling of being a little tense was reported frequently, but only five of thirty-seven described themselves as having a feeling of much tension. Some physical exercise outside work was reported by between 60 and 80% of the workers within the groups. Much of this exercise was connected with work duties at home. In terms of leisure activities, participation in sport was only reported by 12 to 38% in the groups.

The financial and housing situation did not seem to be a stress factor, as indicated in Figure 5.3. Almost all the employees interviewed reported that they could maintain a good domestic economy. Less than 15% of the groups reported difficulties in making ends meet. Their housing situations were assessed as good by all workers except one who described her flat as only satisfactory.

Most subjects found their work satisfying. Their present jobs were preferred by as many as 64 to 100% of the workers within the groups. The remainder wanted to have some other job and only a very few would have preferred to stay at home. This is confirmed by the high proportion of workers within the groups who wanted jobs to obtain social contacts and to improve the family economy.

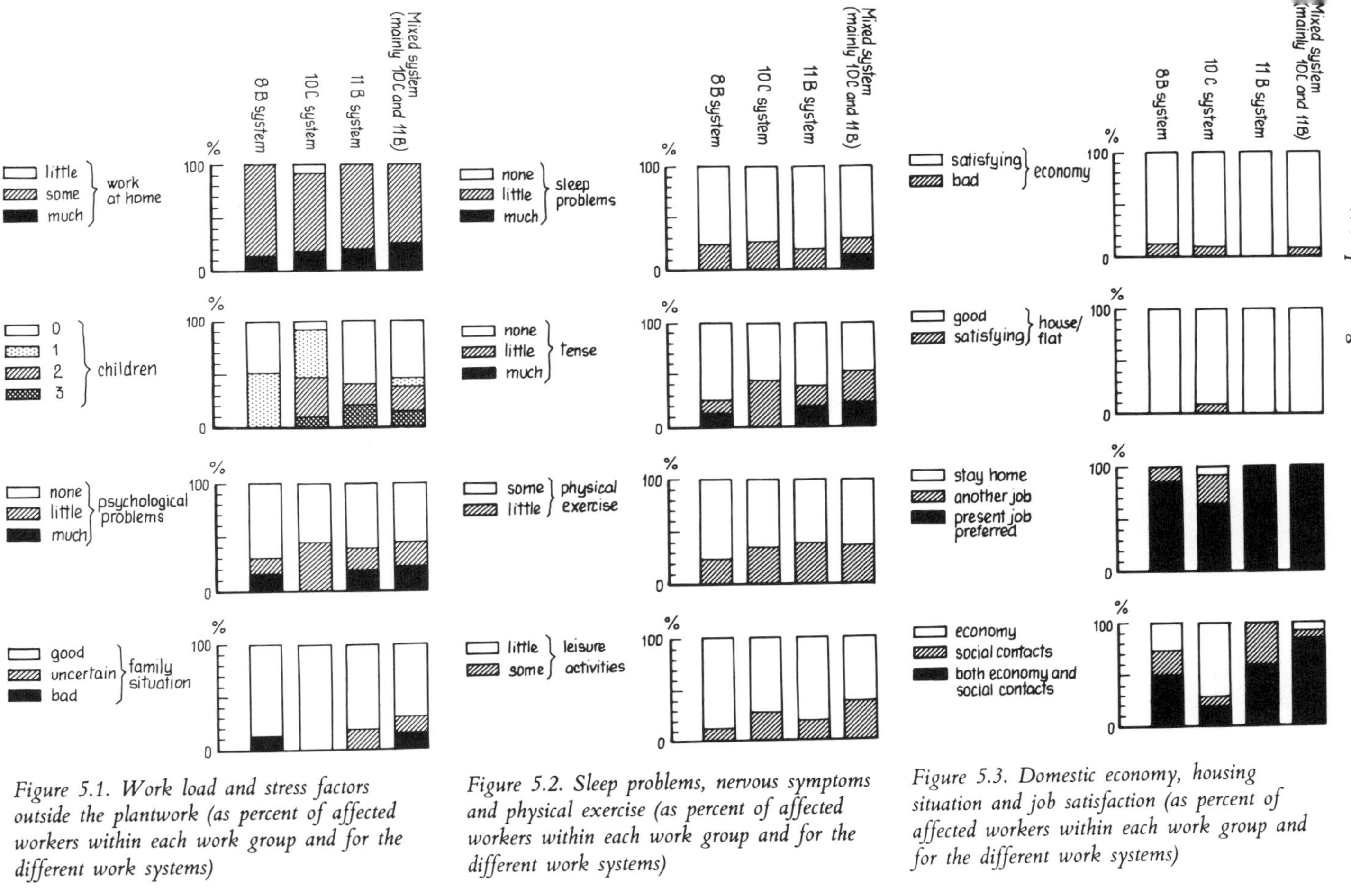

Figure 5.1. Work load and stress factors outside the plantwork (as percent of affected workers within each work group and for the different work systems)

Figure 5.2. Sleep problems, nervous symptoms and physical exercise (as percent of affected workers within each work group and for the different work systems)

Figure 5.3. Domestic economy, housing situation and job satisfaction (as percent of affected workers within each work group and for the different work systems)

Pain level and pain duration

Data on pain intensity experienced during the previous year, during the period of employment and before employment is presented in Figure 5.4.

The body parts most affected were the shoulder/upper arm, neck and low-back. Of the thirty-seven workers, it was found that ten experienced pain intensity at level 3 ('A good deal of pain. Need pauses') and seven at level 4 ('Severe pain. Difficulty in carrying out work') without taking account of body location. All groups reported higher intensity of pain during their time of employment as compared to the previous year before the interview. This difference may be due to many of the employees having worked for nine years previously at unergonomically designed work places, which created a higher trapezius load compared to the present ergonomically designed work station.

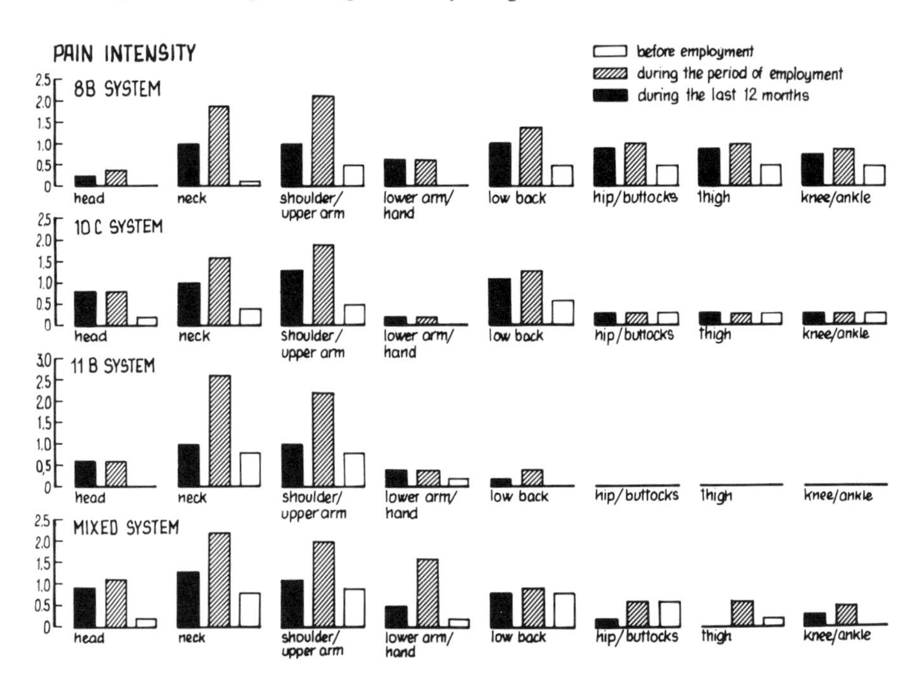

Figure 5.4. Pain intensity for different body areas and work systems

The pain intensity experienced before recruitment was, for most of the groups, much lower when compared to that in the period of employment at the plant, particularly shown by intensity of pain in the upper part of the body. Between 20 and 38% within the groups had suffered pain in some body location before employment at the plant. At their present jobs as many as 63 to 80% within the groups reported pain at work.

Figure 5.5 shows the reports of duration of pain at level 3 or 4 in different body areas during the year before the interview. The proportion of all workers who suffered pain duration of more than thirty days was 19% for shoulder/upper arm, 16% for neck and 11% for low-back. There was no statistically significant difference between the groups.

Between 75 and 85% of the workers within the groups reported the pain to be work-dependent and almost all of them defined a cause at work as a reason for the pain. However, very few took pain-killers, and very few workers needed physiotherapy treat-

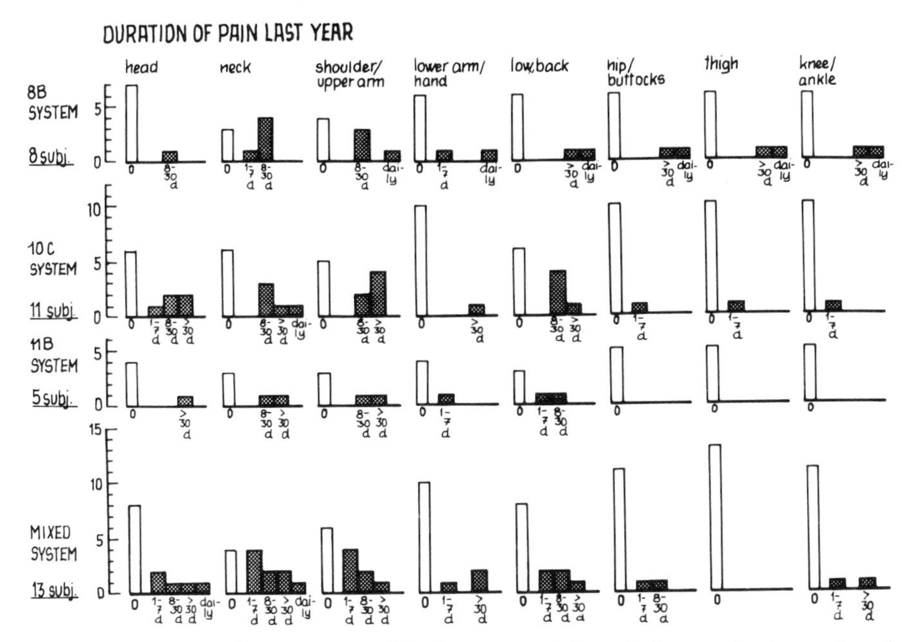

Figure 5.5. Duration of pain intensity of level 3 or 4 in different body areas (as the number of workers for different work systems)

ment during the year before the interview. Of those who had treatment, between 16 and 56% within the groups reported a long lasting positive effect.

Clinical symptoms and signs of musculo-skeletal illness

Clinical signs of musculo-skeletal illness were found in terms of trigger points (pain-provoking points) in the trapezius muscles, restriction of movements in cervical spine and gleno-humeral joints, and pain or tenderness provoked by isometric and endurance tests, as well as when palpating tendon attachments. These signs are shown in Figures 5.6–5.9.

Trigger points

Pain-provoking points were common among workers at the 8B system, but there was no statistically significant difference between the systems.

Restriction of movements

There were no signs of restriction of flexion or extension of the neck, so that all subjects had a range of movement of 45° in both flexion and extension. Only a few workers in some of the work systems showed restriction in the sideways movement of the neck. Rotation of the head exceeded 60° on both sides for subjects on all work systems. However, pain during movements of the neck and/or palpable spasm of the neck muscles were a characteristic feature among 20 to 50% of the workers at the different systems.

The range of movement of the upper arm in the gleno-humeral joint with fixed scapula (passive movements) was normal for both flexion and abduction (i.e. >65°). The same was also true for outwards and inwards rotation in the same joint (i.e. >60°).

Pain or tenderness provoked by isometric and endurance tests showed that of 23 to

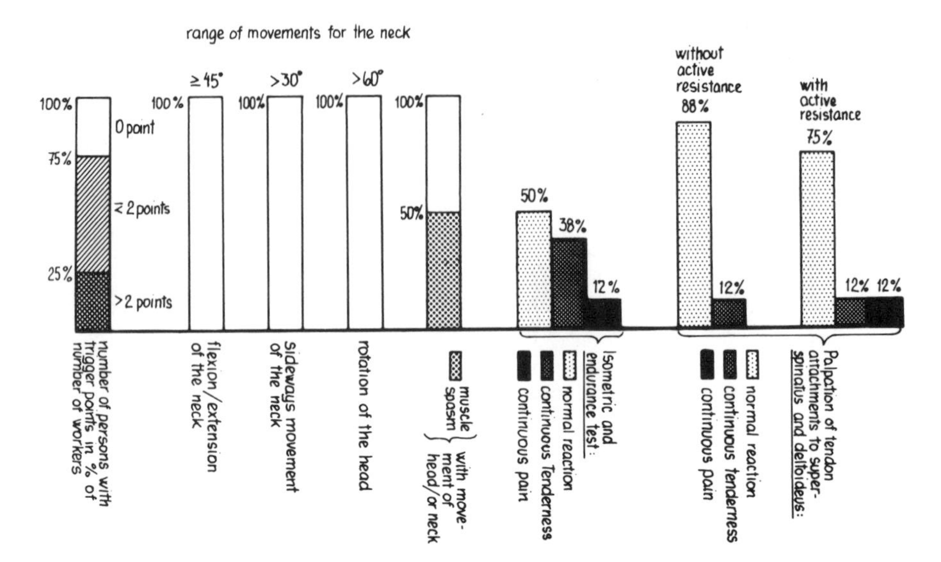

Figure 5.6. Clinical symptoms and signs of musculo-skeletal illness in the upper part of the body for workers on the 8B system

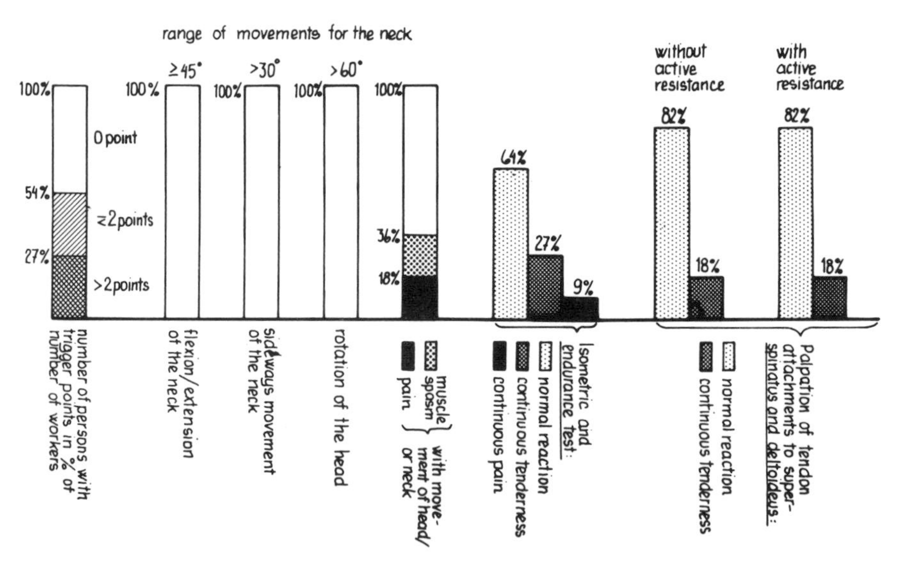

Figure 5.7. Clinical symptoms and signs of musculo-skeletal illness in the upper part of the body for workers on the 10C system

50% of the workers within the groups reported continuous tenderness or continuous pain after the tests. The differences between groups were not statistically significant.

Tendon attachment for m. supraspinatus and m. deltoideus were palpated with or without active muscle contraction. Reports of tenderness or pain varied from nil to 24% of workers in the different groups, but the differences were not significant. Few other

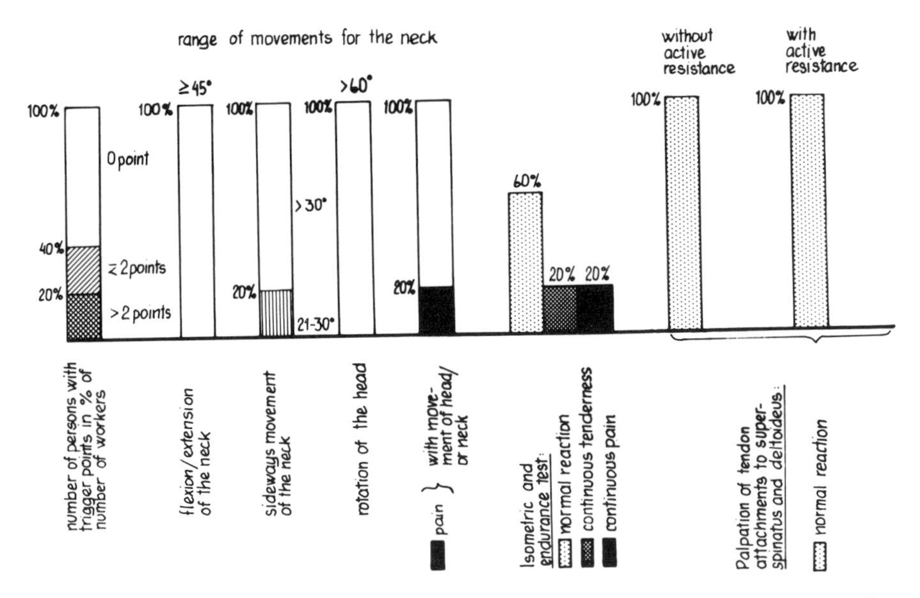

Figure 5.8. Clinical symptoms and signs of musculo-skeletal illness in the upper part of the body for workers on the 11B system

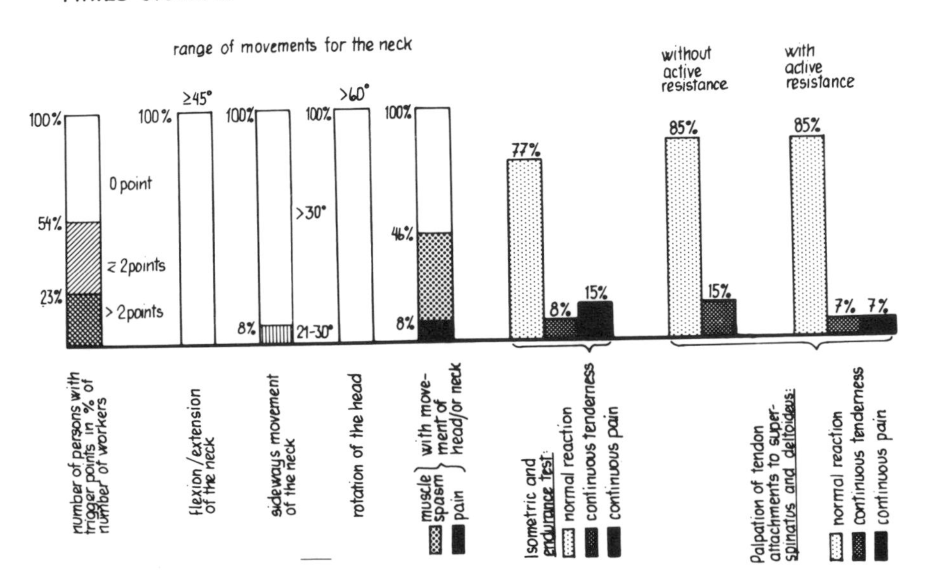

Figure 5.9. Clinical symptoms and signs of musculo-skeletal illness in the upper part of the body for workers on the mixed systems

clinical or neurological signs of pathology were found during the examination.

Clinical examination of the low-back

In all groups, very few subjects showed diminished movement of the low back. The same was also true for the number of subjects who reported pain at full excursion of the movements or pain which started at the beginning of movement. Less than 10% of the workers reported that maximal flexion of the back was restricted by pain (Schobert test). About 10% of the workers had muscle spasm or showed scoliosis in the lumbar back. Other clinical signs from the low-back were few. Very few subjects could give a cause for their low-back pain. Heavy lifting, draughts, cold and wet climate were mentioned as causes.

Discussion

Postural load at work and work duties outside the plant, as well as psychological problems, were all thought to be important factors contributing to the load on the musculo-skeletal system of thirty-seven female workers in this investigation. The effects of postural load and duration of low activity levels ('rest periods') during work had already been shown in previous studies. This study has shown that work at home and serious psychological problems added to the load on the musculo-skeletal system for only a small proportion of the workers. Between 13 and 23% of the workers within each group reported a great deal of work at home or serious psychological problems. However, only two of thirty-seven workers reported both a significant amount of home work and serious psychological problems. A combination of serious psychological problems and some work load at home was also described by three workers. This supports the observation that few workers described themselves as having a feeling of being too tense, and sleep disturbances were only occasionally reported.

Other psycho-social factors such as the family situation, domestic economy and housing situation were reported to create little stress except for a few workers. Factors which might strengthen the musculo-skeletal system, such as sport and other recreational and leisure activities, affected less than 35% of the workers.

Some of the psycho-social questions were sensitive, but all subjects cooperated during the interviews, as their answers were treated confidentially. This was also supported by the level of reproducibility with less than 5% difference between responses at two successive interviews.

Health consequences were evaluated in terms of pain intensity and duration as well as by clinical symptoms and signs of musculo-skeletal illness. Pain intensity during the year before the interview was reported by 27% of all subjects at level 3 (i.e. a good deal of pain and need for pauses) and by 19% at level 4 (i.e. severe pain which did not disappear during pauses and difficulty in carrying out work). In fact, more than 20% of the workers suffered such pain with a duration of more than thirty days in the year before the interview, although very few workers had taken pain-killers or had needed physiotherapy treatment in the same period. More than 75% of the workers reported the pain to be work-dependent, but surprisingly almost the same percentage of the workers preferred to stay in the present job, which they mostly found satisfying. Reasons for this may include strong wishes for social contacts at work and/or improved domestic economy.

Clinical examination may support and give additional information on the subjective experience of pain in evaluation of the prevalence of musculo-skeletal illness. Such illnesses, including tendinitis, myalgia, myotendinitis, lumbago and sciatica, are often clinically associated with muscle tenderness, spasm, hardenings and sore spots when palpating the muscle. This is however a considerable subjective confounding component in such examinations (Waris *et al.*, 1977). It is difficult to confirm local tenderness at tendon and its attachments as a sign of tendinitis. More objective signs of such illness are local swelling of tendon and crepitation during movements of the tendon within the sheath.

Loss of passive movement in joints, together with painful movement, may be measured in a more accurate way. With these limitations of the clinical examination in mind, hardenings and sore spots or 'trigger points' were found in the trapezius muscle for more than 50% of the subjects, as shown in Figures 5.6-5.9.

Movements of the neck were seldom restricted, although spasm of the muscles and/or pain during movement was reported by about 50% of the workers. The isometric and endurance test results showed that 35% of the subjects reported continuous tenderness or pain after cessation of the test. The endurance tests normally produced the highest discomfort or pain level after the tests. These tests were quite representative of the low degree of muscle load which the workers experienced in the real work situations.

Clinical signs of tendinitis were reported by 16% of the subjects when palpating the tendon attachments of m. supraspinatus and deltoideus during muscle contraction against active resistance. This level of tenderness or pain was higher than that reported in tests without active resistance of the movements.

Thus, a substantial number of the subjects reported and showed symptoms and signs of musculo-skeletal illness in the upper part of the body. In fact, such clinical symptoms and signs were found without exception for those who reported pain during work in the period of examination. Furthermore, these tests showed that those who suffered musculo-skeletal illness had reduced ability to tolerate physical load on the musculo-skeletal system.

Clinical examination of the low-back showed that very few had diminished movement of the lumber back or pain during such movement. This is also in agreement with the results from the sick-leave statistics, showing that musculo-skeletal illnesses were much more common in the upper part of the body than in the low-back (Aarås *et al.*, 1988).

Conclusions

The results of the studies have shown that the tasks created low postural load on the workers, measured in terms of trapezius activity and postural angles in the gleno-humeral joint when working at ergonomically designed work places. Less than a quarter of the workers reported other factors which might be considered to influence the prevalence of musculo-skeletal complaints, such as work load outside the plant or serious psychological problems. Nevertheless, serious pain (both in terms of intensity and duration) had been experienced by about a fifth of the workers during the year before the interviews. This high frequency of reported musculo-skeletal complaints was supported by the high prevalence of symptoms and signs of musculo-skeletal illness found at the clinical examination. This indicated that the load on the musculo-skeletal system was higher than should be acceptable, and that a major ergonomics contribution seems crucial to the design of both the work task and the work place, in order to reduce the musculo-skeletal complaints of the workers to an acceptable level. In such design/redesign, job rotation and variety or work posture and movements must be considered.

References

Aarås, A. (1987). Postural load and the development of musculo-skeletal illness. *Scandinavian Journal of Rehabilitation Medicine Supplement* **18**.

Aarås, A. (1989). Acceptable muscle load on the neck and shoulder regions assessed in relation to the incidence of musculo-skeletal sick-leave. Accepted for publication in *International Journal of Human Computer Interaction*.

Aarås, A. and Westgaard, R.H. (1987). Further studies of postural load and musculo-skeletal injuries of workers at an electro-mechanical assembly plant. *Applied Ergonomics*, **18**, 3, 211–219.

Aarås, A., Westgaard, R.H. and Stranden, E. (1988). Postural angles as an indication of postural load and muscular injury in occupational work situations. *Ergonomics*, **31**, 6, 915–933.

Aarås, A., Westgaard, R.H., Stranden, E. and Larsen, S. (1990). Postural load and the incidence of musculo-skeletal illness. In: *Promoting Health and Productivity in the Computerized Office* edited by Sauter, S.L., Dainoff, M. and Smith, M., (London: Taylor & Francis, in press).

Ericson, B.E. and Hagberg, M. (1978). EMG signal level versus external force: a methodological study on computer aided analysis. In: *Biomechanics VI-A*, edited by Asmussen, E. and Jørgensen, K., pp 251–255. (Baltimore: University Park Press).

Jensen, M.P., Karoly, P. and Braver, S. (1986). The measurement of clinical pain intensity. *Pain*, **27**, 117–126.

Kuorinka, I., Jonsson, B., Kilbom, Å., Winterberg, H., Baering-Sørensen, F., Andersson, G. and Jørgensen, K. (1987). Standardised Nordic questionnaires for the analysis of musculo-skeletal symptoms. *Applied Ergonomics*, **18**, 3, 233–237.

Kvarnstrøm, S. (1983). Førekomst av muskel-og skjelettsjukdommer i en verkstadsindustri med saerskild uppmärksamhet på arbeidsbetingande skulderbesvaer. *Arbete och Halsa* **38**. Arbetarskyddsstyrrelsen, Solna, Sweden. (In Swedish).

Larsen, S., Osnes, M. and Lillevold, P.E. (1985). Assessing soft data in clinical trials. Medical Department, CIBA-GEIGY Phara A/S, Stroemmen, Norway.

Sheon, R.P., Maskowitz, R.W. and Goldberg V.M. (1982). *Soft Tissue Rheumatic Pain: Recognition, Management, Prevention*. (Philadelphia: Lea & Febiger).

Waris, P., Kuorinka, I., Kurppa, K., Luopajarvi, T., Virolainen, M., Pesonen, K., Nummi, J. and Kukkonen, R. (1977). Epidemiologic screening of occupational neck and upper limb disorders: methods and criteria. *Scandinavian Journal of Work, Environment & Health*, **5**, Supplement 3, 25–38.

Westgaard, R.H. and Aarås, A. (1984). Postural muscle strain as a causal factor in the development of musculo-skeletal illness. *Applied Ergonomics*, **15**, 3, 162–174.

Westgaard, R.H., Jansen, T.H. and Aarås, A. (1982). Development of musculo-skeletal illness among female workers in work situations requiring static muscle load. Institute of Work Physiology, Oslo. (In Norwegian).

Part III

Physical and mental components of work

The physical and mental components of work are apparent in a group of studies which looked at particular aspects of work rather than complete jobs. However, as these clearly show, the aspects are interrelated and an ergonomic evaluation needs to be made in the full context of the job and cannot simply consider the aspects in isolation.

It can be seen that one of the main preoccupations of ergonomists in industry at present is reducing and controlling the problem of upper limb strain injuries. It is interesting to see the variety of ergonomic techniques which are used by researchers in this field, and the papers illustrate how these can be selected and combined to give information on the different aspects of the demands of the task.

The first paper by Armstrong, Ulin and Ways discusses the use of hand tools, and presents a methodology for job analysis which, as the authors suggest, could be extended to selection of other types of equipment such as jigs, scanners or keyboards. This is based on the use of epidemiological data to identify the jobs requiring further study, which can then be evaluated by psychophysical studies. Cederqvist, Lindberg, Magnusson and Örtengren were faced with a similar question in choosing the appropriate design of screwdrivers for electrical installation work in the construction industry. Their study demonstrates the importance of matching the tool to the workpiece as well as to the workplace.

Tuvesson and her colleagues studied the use of arm suspension balancers which are often suggested as means of reducing the postural load imposed on keyboard operators, but additionally took into account the influence of the mental aspects of the task. Their paper raises the problems of individual differences and preferences which can often be encountered in work design.

Wells, Moore and Ranney tackle the mechanisms of injury causation and propose methods for evaluating the internal loads in the wrist in relation to the main physical risk factors of repetition, posture and force. They have developed instrumentation and a model of the hand and wrist, and demonstrate their application in assessing the consequences of postural differences in terms of muscle loads.

In the final paper of this section, Plette takes an overall look at health and safety in the workplace, reviewing the very many physical, mental and social factors which influence the problem. He sets this in the context of the interdependent nature of the person and the working environment, emphasising the need to balance the level of work stress so that it remains within tolerable limits and within the individual's capacity to adapt.

Chapter 6
Hand tools and control of cumulative trauma disorders of the upper limb

T.J. Armstrong, S. Ulin and C. Ways

Department of Industrial and Operations Engineering, University of Michigan, Ann Arbor, MI 48105, U.S.A.

Introduction

This paper is concerned with the cause and control of upper limb cumulative trauma disorders among workers using hand tools. Examples of these disorders include carpal tunnel syndrome and hand and wrist tendinitis. The causes of cumulative trauma disorders can be classified as personal or work related (Armstrong and Silverstein, 1987; WHO, 1985). Commonly cited personal factors of cumulative trauma disorders include: chronic diseases, acute injuries, congenital defects, aging and gender. Commonly cited occupational factors include: repeated or sustained exertions, forceful exertions, certain postures, localized mechanical stresses, exposure to low temperatures, and hand–arm vibration. The relative contribution of one cause versus another varies from one population to another due to sample selection biases (Monson, 1980; Armstrong and Silverstein, 1987). For example, chronic diseases may be a more significant factor of carpal tunnel syndrome in a clinical population than in a work population, while postural stresses may be a more significant factor for a work population.

Employers are often interested in ascertaining the relative contribution of personal versus work factors for insurance purposes and for prevention of additional cases. Determination of the most important causal factors requires epidemiological studies of injury patterns (Monson, 1980). Injury patterns may be based on compensation reports, plant medical visits, personal medical visits, or worker surveys; however, the number of cases reported under each of these categories may vary considerably from site to site (Fine *et al.* 1986). Unfortunately, there are often too few cases and workers, or the work populations are too unstable, for rigorous statistical analysis. In these instances it may still be possible to identify jobs and populations with elevated incidence rates. In all instances it is desirable that the affected jobs and workers be evaluated to identify recognized cumulative trauma disorder risk factors.

An example of an epidemiological study by Armstrong *et al.* (1987) and Silverstein *et al.* (1987) comparing work and personal factors is shown in Table 6.1. In this study, workers were selected on the basis of their jobs. It was found that the odds of tendinitis and carpal tunnel syndrome (CTS) were 29.4 and 14.3 times greater in workers doing highly repetitive and forceful work than in those doing low repetitive and low force work. The odds of tendinitis in women were 4.3 times that for males. These results can be generalized to provide insight into possible causes of cumulative trauma disorders at other sites.

Table 6.1. Occupational factors versus non-occupational factors

Jobs	n	Tendinitis[a]	CTS[b]
High repetitiveness and high forcefulness	157	10.8%	5.6%
Low repetitiveness and low forcefulness	157	0.6%	0.6%
Plant adjusted odds ratio		29.4[c]	14.3c
Female to male ratio		4.3[c]	1.2

[a] Armstrong *et al.* (1987)
[b] Silverstein *et al.* (1987)
[c] sig.p<.01

A second example from Cannon *et al.* (1981) in which workers were selected on the basis of reported carpal tunnel syndrome is shown in Table 6.2. The odds were found to be 13.8 times higher that persons with carpal tunnel syndrome used vibrating hand tools than control subjects without carpal tunnel syndrome (see Table 6.2). The odds were found to be 12.6 times greater that persons with carpal tunnel syndrome had undergone gynecological surgery than the control subjects.

Table 6.2. 30 (3 males and 27 females) workers with CTS. 90 sex matched controls (from Cannon et al. *(1981)*

	Cases	Controls	Odds Ratio
Gynecological surgery	81.5	25.9	12.6[a]
Vibrating hand tools	70.0	14.4	13.8[a]

[a] sig.p⩽.05

A third study of incidence rates by Armstrong *et al.* (1981) is shown in Table 6.3. These data show that the number and incidence rate of tendinitis, carpal tunnel syndrome and strains range from 8.1 to 13.6 cases per 200,000 hours in the manufacturing departments. The plant-wide average of 5.6 cases per 200,000 hours is very close to the 6.6 value reported by Hymovich and Lindholm (1966) for a six-year period in an electronics plant with 160 workers. Table 6.3 also shows that the relative likelihood of persons reporting these problems in all manufacturing departments is from 7.4 to 12.4 times that of other departments. Based on these data it can be argued that the work activities and equipment in these departments merit further study. Further subdivisions may be made for other job, tool, and worker categories.

In all instances, affected workers should be examined by qualified medical personnel to identify and treat possible personal factors and the associated jobs should be analyzed to identify and correct possible work factors. Job analysis entails: 1) documentation, 2) identification of recognized risk factors, 3) design of job improvements and 4) evaluation of job improvements. This process has been described in other places (Armstrong, 1986; Armstrong *et al.*, 1986), and will be reviewed here in the context of selected hand tools.

Table 6.3. Crude analysis of tendinitis, carpal tunnel syndrome and strains by department during a sixteen month period in a furniture manufacturing plant (Armstrong et al., *1981)*

Dept.	Work hours	Cases	Rate[a]	Rel. risk[b]
Machine	192,775	11	11.4	10.4
Weld	136,715	6	8.8	8.0
Paint	99,060	4	8.1	7.4
Upholstery	182,886	8	8.8	8.0
Assembly	250,055	17	13.6	12.4
Other	943,568	5	1.1	1.0
Total	**1,805,059**	**51–46**	**5.6**	**5.1**

[a] cases per 200,000 work hours or 100 workers per year
[b] relative risk with respect to other departments

These concepts could be extended to other work equipment and activities, such as jigs and fixtures for assembling parts, keyboards for data entry, and scanners for checkout work.

Hand tools contribute to cumulative trauma disorders to the extent that they contribute to repetitive or sustained exertions, forceful exertions, postural stresses, localized contact stresses, low temperatures and vibration. Their contribution must be assessed in terms of the alternatives, e.g., using another tool or no tool. These comparisons may be based on physical parameters such as weight, torque and work location; they may also be based on psychophysical ratings. By using these comparisons it should be possible to select tools and work designs that minimize the risk of cumulative trauma disorders.

Repetitive exertions

Ways of using the same tool vary considerably from job to job. In a study of the vibration exposure in an assembly line of seventeen workers building small electrical appliances, Radwin and Armstrong (1985) found that the powered screwdrivers were operated for between 4.7 and 58.5 minutes in 8 hours of work; in many cases the screwdrivers were held constantly. At one extreme, the screwdriver was used only occasionally to insert screws missed by other workers; at the other extreme, the screwdriver was used repeatedly to drive 1,800 screws per hour. The stresses associated with a tool used only occasionally would have to be very high to result in a cumulative trauma disorder, whereas the stresses associated with a tool used repeatedly could be very low and still result in a cumulative trauma disorder. However, it cannot be stated that the first job is not stressful or will not result in a cumulative trauma disorder; the balance of the job time may be committed to highly stressful materials handling or assembly tasks. If this is the case, tool improvement probably would not reduce the risk of cumulative trauma disorders.

Even when the use of a particular tool results in significant stresses, these stresses may be caused by additional factors such as process specifications, materials, work station layout, work methods and worker attributes. The contribution of these factors can be demonstrated by considering how the job would be performed under alternative conditions. For example, Tichauer (1966) described the highly repetitive use of a ratcheting screwdriver as a factor in carpal tunnel syndrome. One thousand screws per day resulted in 5,000 exertions. By using a powered screwdriver, it is possible to reduce the number of exertions from 5,000 to 1,000 per day. This change can reduce repetitiveness and the risk of carpal tunnel syndrome; also, it is possible to select a powered screwdriver that can improve torque consistency to achieve better work quality. It can also be argued that the work pace could be increased by using the powered tool. By conventional industrial engineering standards, the worker would no longer be working at 100%; the temptation to re-study and re-rate the job would arise. If the worker were on an incentive program, the temptation to go faster to increase earnings would arise. Increasing the pace of the job would probably increase productivity, but the number of exertions and the risk of carpal tunnel syndrome would not be reduced in comparison to the case of the manual screwdriver. It is possible that the tool may affect the risk of carpal tunnel syndrome in other ways: there could be increases in force requirements, stressful postures, contact stresses, and exposure to low temperatures and vibration.

Forceful exertions

Force requirements must be considered for both holding a tool and operating it. The force to hold a tool is related to the tool's weight, handle size, shape and friction (Ayoub and LoPresti, 1971; Pheasant and O'Neill, 1975; Drury, 1980; Armstrong, 1986; Buchholz *et al.*, 1988). It can be argued that the tool force requirements and the risk of cumulative trauma disorders can be lowered by selecting tools with low weight, optimally sized handles, and high friction handles. Unfortunately, it is possible that a tool designed to minimize force may not work for some tasks. For example, Radwin and Armstrong (1985) found that some people held the tools continuously, even though the tools were used less then 12% of the time. A handle size optimized for strength may be much too large in this work situation. No matter how much a tool reduces stress, if it interferes with the job or slows down the worker, it will probably not meet with wide acceptance.

The force required to operate a tool can be many times greater than the force required to hold it. For example, a one kilogram tool can easily produce three newton-meters torque. For a given tool it can be argued that the force requirements will be higher for a high torque fastener than for a low torque fastener. For a given fastener application it can be argued that more hand force will be required to use an in-line tool than to use a pistol shaped tool; more force will be required to use a pistol tool than a right angle tool. Which tool is used may be influenced by the geometry of the work object or of the work station. Another consideration is the duration of the torque force. Some tools may achieve the desired torque very quickly and then automatically disengage or shut off; other tools may require a long time to achieve the desired torque and then maintain that torque until the tool is shut off by the operator (VanBergeijk, 1987). In addition to the tool speed and the torque control mechanism, the force duration will be affected by the length and pitch of the fastener and by the hardness of the material. Torque forces may also be affected by the use of external devices such as torque bars and articulating arms (VanBergeijk, 1987).

Stressful postures

Postural stresses can be predicted based upon biomechanical arguments, laboratory studies and epidemiological studies which indicate that working with the elbow elevated is more stressful than working with the elbow at the side of the body (Armstrong, 1986). Similarly, it can be argued that the elbow should not be highly flexed, the forearm should not be rotated to one extreme or the other, and that the wrist should not be deviated, flexed or hyper-extended (Armstrong and Silverstein, 1987; Armstrong, 1986).

In order to illustrate the postural concerns of tool design, a work situation may be considered in which a powered screwdriver is used to insert screws into a vertical surface. Though this is a hypothetical example, many such jobs can be found in automobile, home appliance, furniture and many other bench assembly operations. Based on a process of elimination, the least stressful posture for driving screws on a vertical surface using a pistol-shaped screwdriver is one where the elbows are at the side of the body and the long axis of the forearm and the tool are horizontal. The ideal work height can be computed as elbow height plus the vertical distance from the center of grip to the center of the bit. The vertical height of the tool varies from one type of tool to another and may be determined by the fastener force and torque requirements. If a great deal of force is required to keep the bit of the tool engaged, it is desirable that the vertical distance be minimized to minimize torque on the wrist. If a high level of torque is required to tighten the screw, it is desirable that the distance be increased to reduce torque reaction forces on the hand and forearm. Often, high torque forces are accompanied by high forces to engage the bit. If this is the case, it may be desirable to deal with the engagement forces through bit and fastener design rather than through tool design.

Once it has been determined that the work is to be performed on a vertical surface and a pistol-shaped tool has been selected, a work location can be selected so that the elbow is at the side of the body and the hand and forearm form a straight line (See Figure 6.1). Elbow height is predicted as 63% of stature plus shoe height (Drillis and Contini, 1966). Average elbow height for a U.S. female who is fifth percentile in stature (151.1 cm) is 95.2 cm; average elbow height for a U.S. male who is ninety-fifth percentile in stature (186.9 cm) is 117.8 cm (USDHEW, 1979). It can be predicted that postural stresses can be minimized for persons wearing two centimeter high shoes and using a 4.4 cm tool by making the work surface adjustable from 101.6 cm to 124.2 cm. In the absence of better information, it is desirable to locate the work in this predicted range; however, ultimately it is desirable to verify the prediction with worker performance and health data. Unfortunately, it would be necessary to isolate a stable population of several hundred people performing this type of work to have a 90% chance of finding 5% prevalence difference with 95% confidence. Seldom is it possible to find such a study population. An alternative is to study perceived exertion for a representative group of workers.

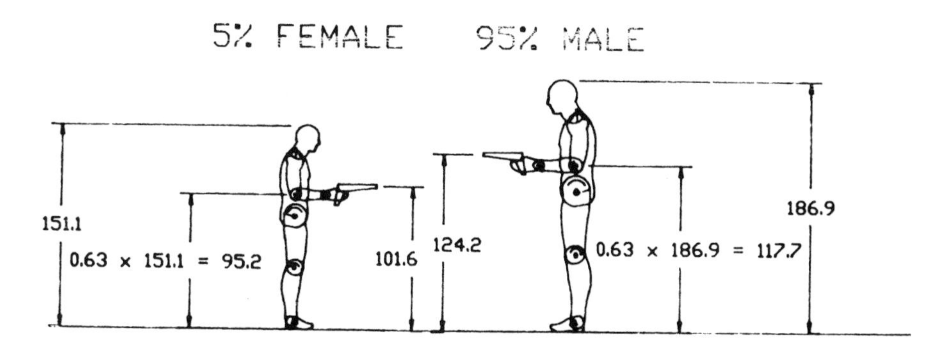

Figure 6.1 Work should be located so that the elbow is at the side of the body and the hand and forearm form a straight line (from Armstrong, 1986)

Both inexperienced and experienced workers have advantages as subjects for evaluating work equipment and procedures. Inexperienced workers are less likely to be biased by established customs and procedures than are experienced workers, but they may lack the skill required to use tools in such a way that they can be critically evaluated. Also, it may be difficult to locate an adequate population of experienced workers of the desired age and gender. Theoretical predictions such as those shown in Figure 6.1 may also be evaluated by studying incumbent workers using similar tools as described by Armstrong *et al.* (1989). Twenty-three workers using a total of thirty-three tools used on an automobile assembly line were asked to rate handle size and weight, tool force and posture comfort of their tools on scales of zero to ten. Subject preferences were then compared with measured physical values. Vertical work locations between 100 cm and 150 cm were significantly more comfortable than were work locations below 100 cm or above 150 cm. Critical levels were also found for horizontal locations, tool mass and handle circumference. The effects of gender and stature on worker ratings were insignificant in comparison with the effects of the tool and work station parameters.

Psychophysical assessment techniques using both inexperienced and experienced subjects become even more attractive for studies of contact stresses, thermal extremes and vibration, where there are not well-developed theoretical arguments for selecting design configurations that minimize risks of cumulative trauma disorders. Such techniques have been applied to the study of vibration.

Although theoretical arguments and psychophysical data can be used to select and design jobs that minimize work stresses, it cannot be said with certainty that persons performing those jobs will not develop carpal tunnel syndrome, hand and wrist tendinitis or other cumulative trauma disorders. It has been shown that the risk of back injuries increases as psychophysical guidelines are exceeded (Snook *et al.*, 1978). The demonstration of such a relationship for hand intensive tasks and cumulative trauma disorders will have to await further studies.

Conclusions

Epidemiological studies are required to determine the relative contribution of personal and work related factors of cumulative trauma disorders. Epidemiological data also provide a basis for determining which jobs merit further study to identify and correct work factors. In all cases, however, affected workers should be examined by qualified medical personnel and associated jobs should be studied. Hand tools may contribute as a work related factor to the extent that they make the task more repetitive or forceful, increase contact and postural stresses, or increase exposure to vibration and low temperatures. The contribution of a given tool can be assessed in terms of how long and how frequently the tool is used and by comparing that use with that of other tools or no tool at all. Comparisons based on physical properties of the tools and theoretical arguments should be evaluated using psychophysical studies. Additional studies are required to demonstrate that tool-tasks selected on the basis of theoretical and psychophysical studies reduce the risk of cumulative trauma disorders.

References

Armstrong, T. (1986). Ergonomics and cumulative trauma disorders. *Hand Clinics*, **2**, 553–566.

Armstrong, T.J. (1986). Upper-extremity posture: definition, measurement and control. In *Ergonomics of Working Postures* edited by N. Corlett, J. Wilson and I. Manenica, (Philadelphia: Taylor and Francis), 59–73.

Armstrong, T.J. and Silverstein, B.A. (1987). Upper-extremity pain in the workplace—role of usage in causality. In *Clinical Concepts in Regional Musculoskeletal Illness*, edited by N. Hadler, (Grune & Stratton, Inc.), 333–354.

Armstrong, T., Foulke, J., Goldstein, S. and Joseph, B. (1981). *Analysis of Cumulative Trauma Disorders and Work Methods*. Project Report RFQ 80–27/PO 80–2891, National Institute for Occupational Safety and Health, Cincinnati, OH.

Armstrong, T.J., Radwin R.G., Hansen, D.J. and Kennedy, K. (1986). Repetitive trauma disorders: Job evaluation and design. *Human Factors*, **28**, 325–336.

Armstrong, T.J., Fine, L.J., Goldstein, S.A., Lifshitz, Y.R. and Silverstein, B.A. (1987). Ergonomic considerations in hand and wrist tendinitis, *Journal of Hand Surgery*, **12A**, 830–837.

Armstrong, T.J., Punnett, L. and Ketner, P. (1989). Subjective worker assessments of power hand tools used in automobile assembly. *American Industrial Hygiene Association Journal*, **50**, 639–645.

Ayoub, M.M. and LoPresti, P. (1971). The determination of an optimum size cylindrical handle by use of electromyography. *Ergonomics*, **14**, 509–518.

Buchholz, B., Frederick, L.J. and Armstrong, T.J. (1988). An investigation of human palmar skin friction and the effects of materials, pinch force and moisture. *Ergonomics*, **31**, 317–325.

Cannon, L., Bernacki, E. and Walter, S. (1981). Personal and occupational factors associated with carpal tunnel syndrome. *Journal of Occupational Medicine*, **23**, 255–258.

Drillis, R. and Contini, R. (1966). *Body Segment Parameters*. Report No. 1166–03, Office of Vocational Rehabilitation, Dept. of Health Education and Welfare, (New York: N.Y. University, School of Eng. and Science).

Drury, C.G. (1980). Handles for manual materials handling. *Applied Ergonomics*, **11**, 35–42.

Fine, L., Silverstein, B.A., Armstrong, T.J., Anderson, C.A., and Sugano, D.S. (1986). Detection of cumulative trauma disorders of upper extremities in the workplace. *Journal of Occupational Medicine*, **28**, 674–678.

Hymovich, L. and Lindholm, M. (1966). Hand, wrist, and forearm injuries—the result of repetitive motion. *Journal of Occupational Medicine*, **8**, 573–577.

Monson, R.R. (1980). *Occupational Epidemiology*. (Boca Raton: CRC Press, Inc.).

Pheasant, S. and O'Neill, D. (1975). Performance in gripping and turning. *Applied Ergonomics*, **6**, 205–208.

Radwin, R.G. and Armstrong, T.J. (1985). Assessment of hand vibration exposure on an assembly line. *American Industrial Hygiene Association Journal*, **46**, 211–219.

Silverstein, B.A., Fine, L.J. and Armstrong, T.J. (1987). Occupational factors and carpal tunnel syndrome. *American Journal of Industrial Medicine*, **11**, 343–358.

Snook, S.H., Campanelli, R.A. and Hart, J.W. (1978). A study of three preventative approaches to low back injury. *Journal of Occupational Medicine*, **20**, 478–481.

Tichauer, E. (1966). Some aspects of stress on forearm and hand in industry. *Journal of Occupational Medicine*, **8**, 63–71.

USDHEW (1979). Weight and Height of Adults 18–74 Years of Age: United States 1971–74. *Vital and Health Statistics* Series Number 211, US Dept. HEW PHS, Office of Health Research, Statistics, and Technology (National Center for Health Statistics, Hyattsville, MD).

VanBergeijk, E. (1987). Selection of power tools and mechanical assists for control of occupational hand and wrist injuries. In *Ergonomic Interventions to Prevent Musculoskeletal Injuries in Industry*, (Chelsea: Lewis Publishers, Inc.), 151–159.

WHO, (1985). *Identification and Control of Work-Related Diseases*. Report of a WHO Expert Committee, WHO Technical Report Series 714, (Geneva: WHO), 9.

Chapter 7
Influence of cordless rechargeable screwdrivers on upper extremity work load in electrical installation work

Tony Cederqvist

Research Foundation for Occupational Safety and Health in the Swedish Construction Industry, S-18211 Danderyd, Sweden

Mats Lindberg, Bo Magnusson and Roland Örtengren

Department of Industrial Ergonomics, Linköping University of Technology, S-58183 Linköping, Sweden

Introduction

Handtools are involved in many injuries which are costly, severe and occur frequently in industry. The upper extremities are injured more often than any other part of the body (Aghazadeh and Mital, 1987). In the Swedish construction industry, about one fifth of all reported accidental injuries are due to the use of man powered or externally powered handtools, accounting for 56% and 32% respectively (Axelsson and Fång, 1985). Over-exertion is the most common factor connected with occupational diseases caused by hand-tools, with 56% being caused by man powered handtools and 35% by externally powered handtools.

Electrical installation work in the Swedish construction industry has changed over the years, from varied mechanical work to monotonous short-cycle assembly activities where manual screwdriving plays a significant part (Ahlgren *et al.*, 1983). Over 80% of the work consists of installation tasks, and more than one third of the working time is spent with the arms elevated above shoulder level (Jarebrant, 1987). Among electricians in Sweden, symptoms such as discomfort and pain are common in the neck, shoulder and arms. Each year approximately one third of the electricians report pain from the elbow region and two thirds from the shoulder (Bjäreborn, 1984; Lindstrand, 1988). Most electricians consider manual screwdriving and drilling as the main cause of pain in the elbow region (Ahlgren *et al.*, 1983; Bjäreborn, 1984). In the assembly industry, frequent use of powered screwdrivers seems to be associated with a high risk of developing pain in the neck, shoulder, forearm and wrist. Here the reason is believed to be repetitive high axial forces with which the driver is pressed against the screw and workpiece, sometimes demanding the use of both arms and even the weight of the body (Arnkvaern, 1985). It has also been shown that higher shut-off torques require significantly higher gripping forces by the operator (Johnson and Childress, 1988). Arnkvaern (1985) showed the importance of the condition of the screwdriver head and the type of screw head used with respect to the axial forces which have to be exerted. Phillips-type screws demanded 36% higher axial force compared to TORX screws when a new Phillips screwdriver head was used, and 111% higher when a worn head was used.

The selection of screwdrivers is very critical because their design influences the workload and the risk of injuries. In construction work, it is not unusual to find electric drills being used as screwdrivers. This leads to transmission of high torques to the hand since there are seldom any releasing clutches or speed regulators. Furthermore, drills are often heavy and, when held in the hand, their centres of gravity are often situated in front of the grip, which leads to a high load on the shoulder and wrist. They also restrict mobility due to the need for a power cord. Other important factors that influence the workload are screw type, hole material, working height, working technique and experience.

A few years ago, small lightweight drills and drivers powered by rechargeable batteries became available in the construction industry. At first, these could not meet users' demands on performance and quality and were regarded as toys among construction workers. However, the design of these new screwdrivers has improved and they have become more powerful, reliable and capable of meeting the demands of professional users. The use of rechargeable screwdrivers has been reported to be beneficial for electricians suffering from epicondylitis, as well as improving the efficiency of the work (Ahlgren *et al.*, 1983). Many electricians have now received screwdrivers from their employers while some have bought them themselves, but there are still many who for various reasons use only the ordinary screwdriver. It has therefore been in the interest of EFAK, the Swedish joint employer-employees working environment committee, to see that more cordless powered screwdrivers are used in electrical installation work. However, it is obvious that the ergonomic consequences of increased use of such screwdrivers may be either good or bad, depending on the tool design and the work task. The aim of the present study has been to investigate the workload when using man powered and cordless powered screwdrivers in typical work situations. A secondary aim has been to study the influence of different types of screwdrivers and screws.

Material and methods

Ten male electrical installation workers, ranging from 19 to 48 years of age, participated in the study. All except one were right-handed and all had experience of using rechargeable screwdrivers. Several operator characteristics were recorded, and grip and shoulder strengths were measured with a force transducer (Bofors KRG-4).

During the experiments, a rechargeable screwdriver (Panasonic EZ 570) and an ordinary screwdriver (Bahco 8155) were used in a simulation of assembly operations performed in electrical installation work. These tools (shown in Figure 7.1) were chosen because they are common and are considered to be good representatives of their respective classes.

The Panasonic EZ 570 is a plastic, pistol-shaped, rechargeable screwdriver that automatically shuts off at a preset torque. The handle diameter is 4.5 cm and the weight 1.6 kg. The screwdriver was run at 350 rpm and the preset shut-off torque was 2.5 Nm. The Bahco is a modern screwdriver with a plastic handle and is stated to be designed in accordance with ergonomic criteria. Handle diameter is 3.2 cm and weight 0.175 kg. Screws with Phillips and TORX heads were used, as shown in Figure 7.2.

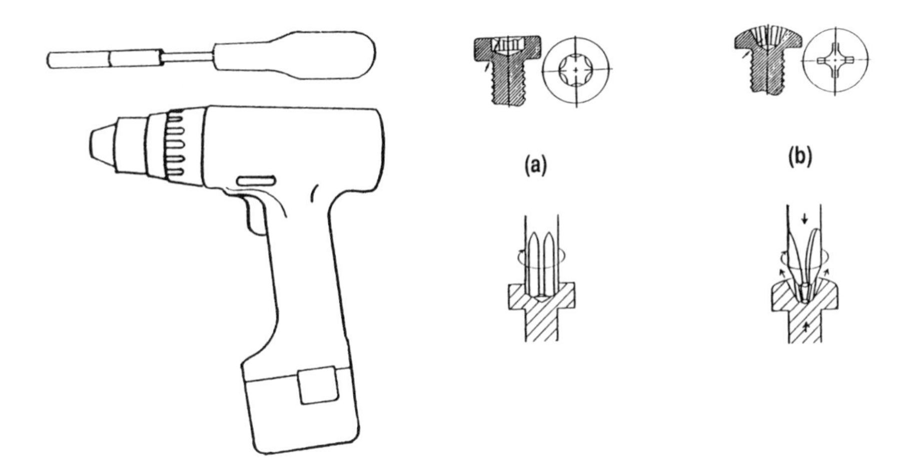

Figure 7.1. The screwdrivers. The Bahco ergo 8155 screwdriver and the Panasonic EZ570 rechargeable drill and driver.

Figure 7.2. Sketches of the screw types. TORX (a) and Phillips (b).

The work was performed in a specially designed test-rig (shown in Figure 7.3) with a workpiece attached. The workpiece was a wooden beam measuring 45mm × 95mm × 800mm. The test-rig could be arranged in different positions so that both working height and orientation of the workpiece could be varied. A load cell (Shinkoh U2D1-50K) was attached to the workpiece so that axial pressure on it could be measured.

The subjects took up a standing position in the work area, which was defined by markings on the workpiece. The task consisted of inserting Phillips and TORX screws (4.8mm × 19mm) into the wooden beam with the aid of the two screwdrivers. There was a three-minute working session for each combination. Phillips No. 2 and TORX T25 screwdriver heads were used. The work was performed at three different standardized working heights which will be referred to in the following as, above head level (AHL), eye level (EL) and hip level (HL). The screwing directions were upwards, horizontal and downwards, respectively. The order in which the sessions were performed was randomized to minimize any sequential effects. The subjects were allowed to choose their own pace, which was recorded.

To measure the muscular load, myoelectric signals (MES) were recorded by means of bipolar surface electrodes from the *extensor carpi radialis brevis* (ECRB), the *flexor digitorum superficialis* (FDS) and the *trapezius pars descendens* (TRAP) of the dominant arm. The signals were amplified by means of small preamplifiers (TB Elektronik) fastened to the skin near the electrodes. The preamplifiers were connected to main amplifiers (TB Elektronik) by means of a 10 meter long cable, allowing the subjects to move freely. The MES were monitored on an oscilloscope screen and the root-mean-square (rms) signals were tape recorded (TEAC R-71) for subsequent analysis. The MES were then fed to a personal computer (Victor V286) and digitized at a sampling rate of 25 Hz. Using a signal analysis program system ARNESE (Lindberg, 1987) the 10%, 50% and 90% levels of the amplitude distribution of the rms detected MES were calculated.

For calibration and normalization purposes, the relationship between voluntary contraction force (MVC) and MES amplitude (MVE) up to maximum effort was established before

Figure 7.3. Sketches of the working heights. Above head level (AHL), eye level (EL) and hip level (HL)

each session for each muscle, using weights and force transducers. Thus, the muscle load could be expressed either as the actual force in relation to maximum voluntary force (%MVC) or as MES amplitude in relation to maximum voluntary signal amplitude (%MVE).

The applied axial force with which the screws were pressed against the wooden beam was measured with the force transducer and also recorded on tape for later evaluation on the computer, where peaks and time patterns could be studied. After each session, the subjects were asked to rate the level of perceived exertion according to the Borg scale (Borg, 1980).

Differences in results between experimental conditions were tested statistically by means of paired t-tests (two-tailed). Tables of results and statistical tests can be obtained from the authors.

Results

The results are presented in Figures 7.4 to 7.7. For the ECRB muscle, there were statistically significantly lower peak loads and median loads when using the powered screwdriver than when using the manual screwdriver, irrespective of working height and screw type. The static component (10% ADL) was significantly higher when working at eye and hip level for TORX and at hip level for Phillips, but in other positions it was lower (Figure 7.4). The picture was similar for the FDS muscle (Figure 7.5). The loading pattern shown by the trapezius muscle was different. Here the load level differed more between screw types than between manual and machine power, (Figure 7.6) and the peak and the median loads were higher for machine screwing above head level and at eye level than for manual screwing, although the static component was lower. When working at hip level, machine screwing showed lower peak values while the static component was higher than in manual screwing.

For all working heights the ratings of perceived exertion were lower for machine screwing than for manual screwing with the same screw type, but the differences were not always statistically significant (Figure 7.7). TORX screws on the other hand, always had statistically significantly lower ratings than Phillips screws. Also axial force was always lower for TORX screws than for Phillips screws. Machine screwing with TORX required the lowest axial forces at all heights, and manual screwing with Phillips required the highest. There was no significant difference in pace between TORX and Phillips screws. However, the powered screw driver was used at a significantly higher pace at all working heights compared with the manual screwdriver.

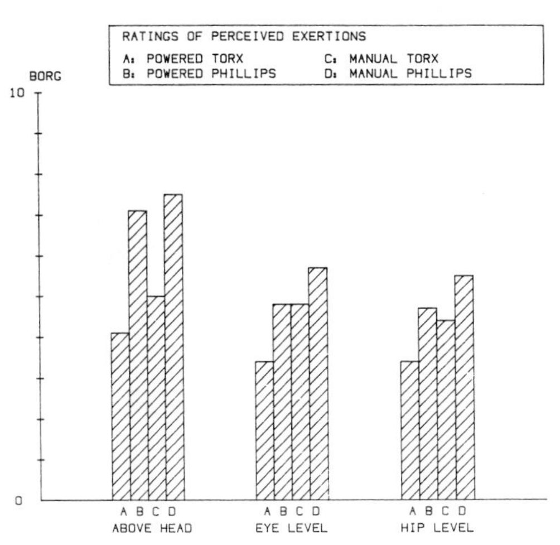

Figure 7.7. Ratings of perceived exertion for the different sessions.

Discussion

One aim of this paper was to investigate muscular load levels in typical electrical installation assembly tasks when inserting screws by means of powered and manual screwdrivers. The results show that peak and median amplitude load levels (90% and 50% ADL) in the ECRB and FDS muscles decreased substantially when the powered screwdriver was used as compared to the manual screwdriver, at all working heights for both TORX and Phillips screws. The average peak levels were up to 65% and 80% MVE for ECRB and FDS muscles respectively when using the manual screwdriver and up to 26% and 35% MVE for the powered screwdriver. Average mean amplitude levels of up to 21% and 40% MVE were found for ECRB and FDS muscles respectively with the manual screwdriver compared to 13% and 18% MVE for the powered screwdriver. The static load levels (10% ADL) were about the same irrespective of which screwdriver and screw type was used, the differences being in most cases no more than 2 to 3% MVE. However, the magnitude of static load in the forearm exceeded 5% MVE and was sometimes up to 14% MVE.

For the trapezius muscle, the load pattern was different. For both peak and mean load

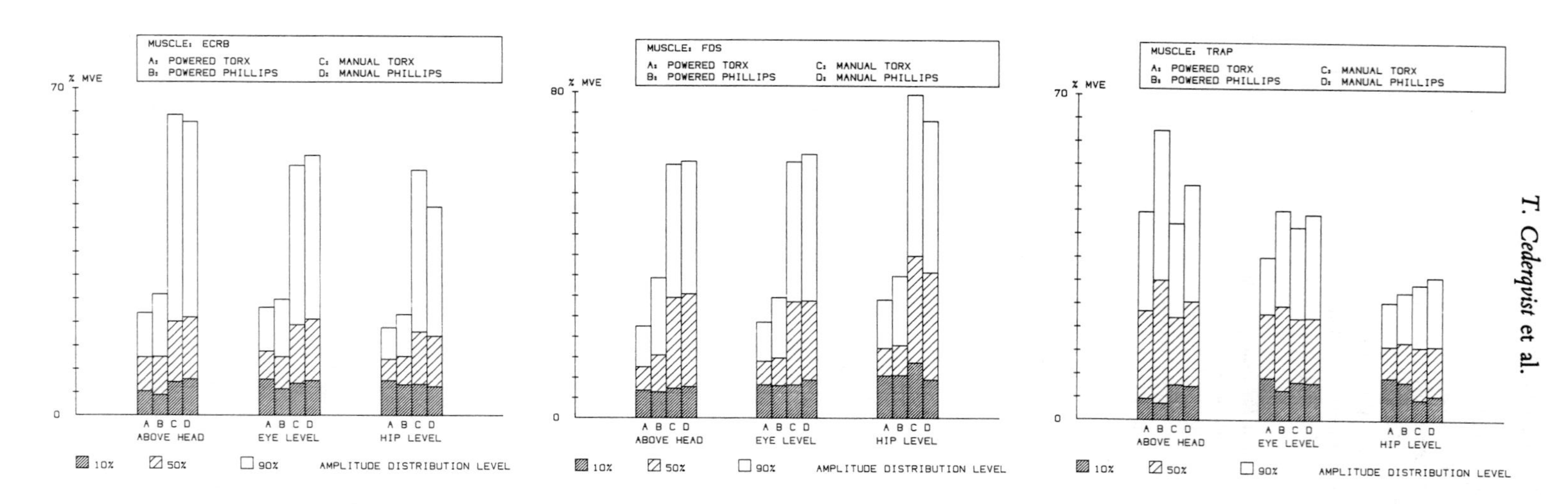

Figure 7.4. Amplitude distribution levels for the ECRB muscle.

Figure 7.5. Amplitude distribution levels for the FDS muscle.

Figure 7.6. Amplitude distribution levels for the trapezius muscle.

levels, the working height seemed to be the critical factor as a decrease in height resulted in decreased amplitude levels. Peak load levels were up to 62% MVE when working above head level, while at hip level they were up to 30% MVE. Mean load levels were up to 30% MVE when working above head level, while at hip level they were about 15% MVE and about the same for both screwdrivers and screw types. The static load levels varied between 5% and 10% MVE. Static load levels were similar for the screwdrivers at eye level, while at hip level and above head level there were significant differences between screwdriver types.

The results indicated that the heavy, powered screwdriver was handled with less static load level than the ordinary screwdriver when working above head level, while the opposite was found at hip level. This finding shows that the weight of the tools has an important influence on the preferred working technique. When the perceived muscle load is high because of tool weight, screw changing results in an automatic lowering of the arm to reduce the static load. When inserting screws in a downward direction on a horizontal surface, the perceived weight of the tool is markedly reduced and thus a more monotonous working technique can be tolerated.

Ratings of perceived exertion were between 3 and 7 on the Borg scale, which is rather high. Differences between experimental conditions show that the subjects preferred the powered screwdriver/TORX combination. The pattern of ratings was fairly similar to the trapezius muscle loading pattern. Also the axial force measurements showed a pattern similar to the ratings of perceived exertion. Highest overall axial forces (up to 140 N) were found for the manual screwdriver/Phillips screw combination at hip level. The lowest axial forces were found for the powered screwdriver/TORX combination at eye level and above head level, where they never exceeded 70 N. Second best was the manual screwdriver/TORX combination reflecting the importance of the selection of screw head types in accomplishing a reduced axial force. None of the subjects had any previous experience of TORX screws, which may have caused them to apply a higher axial force than necessary. If so, the force could be further reduced.

A higher working pace was found when inserting screws with the powered screwdriver, especially at low working heights. The highest pace was found at hip level with 11 screws per minute when using the powered screwdriver. There was also a tendency towards a higher working pace when the powered screwdriver/TORX combination was used at all working heights. However, the effect was not statistically significant.

It has been shown that shoulder muscles in particular are frequently affected by work-related conditions in a variety of jobs. The reason is believed to be high static load levels, often accompanied by a fairly low mean level (Jonsson, 1988). In construction work, the picture is to some extent different with respect to muscular load levels. Although jobs have become more monotonous and assembly-like (Ahlgren *et al.*, 1983), which tends to give higher static load levels, it has been argued that most jobs still require high peak loads. This study confirms that both peak loads and static loads may be high in construction work.

All the muscle load values mentioned above exceed the threshold values suggested by Jonsson (1978a, b). This means that the work load in the typical electrical installation work situations simulated in this study is probably high enough to cause overload disorders and pain. The use of powered screwdrivers, however, can significantly reduce the peak and mean load levels in lower arm muscles and tolerable load levels are reached, especially in combination with TORX screws. The trapezius muscle shows a more complex load pattern, indicating that working height and tool weight must be considered as well. As the static load also depends on the time pattern of the work, it is important to remember not to convert the benefit of a reduced load level into a prolonged working time.

Conclusions

The present study shows that the work load when inserting screws with a manual screwdriver is high enough to cause symptoms of overload disorders. Using a powered screwdriver reduces the load in the forearm muscles to more favourable levels at all working heights, while in the trapezius muscle this is not always the case. Here the weight of the tool and an unfavourable working height have a greater influence on the work load. The use of powered screwdrivers increases the work pace. Screw design also influences the work load. TORX type screws require less axial force than the Phillips type. Although the work performed in this study was simulated installation work, the similarity to real working conditions is high enough for increased use of powered screwdrivers to be recommended as one way of reducing overload disorders.

References

Aghazadeh, F. and Mital, A. (1987). Injuries due to handtools. *Applied Ergonomics*, **18.4**, 273–278.

Ahlgren, Å., Magnusson, S., Thorson, J. and Åkesson, G. (1983). *Pains in the arm due to screwdriving*. Bygghälsan (in Swedish).

Arnkvaern, K. (1985). *Axial force requirements when inserting TORX and Phillips type screws*. Course project, The Swedish National Board of Occupational Safety and Health (in Swedish).

Axelsson, P.-O. and Fång, G. (1985). *Accidents with handheld machinery and tools in the construction industry*. Occupational Accident Research Unit, Royal Institute of Technology (in Swedish).

Bjäreborn, B.-M. (1984). *Prevalence of elbow disorders in electrical installation work*. Course project, The Swedish National Board of Occupational Safety and Health (in Swedish).

Borg, G. (1980). A category scale with ratio properties for intermodal and inter-individual comparisons. In: *Proceedings of XXIInd International Congress of Psychology*, Leipzig, GDR, July.

Jarebrant, C. (1987). *Work breaks in electrical installation work*. Course project, Lund University (in Swedish).

Johnson, S.L. and Childress, L.J. (1988). Powered screwdriver design and use: tool, task, and operator effects. *International Journal of Industrial Ergonomics*, **2**, 183–191.

Jonsson, B. (1978a). Kinesiology—with special reference to electromyographic kinesiology. *Contemporary Clinical Neurophysiology* (EEE Suppl. No. **34**), 417–428.

Jonsson, B. (1978b). Quantitative electromyographic evaluation of muscular load during work. *Scandinavian Journal of Rehabilitation Medicine*, Suppl. **6**, 69–74.

Jonsson, B. (1988). The static load component in muscle work. *European Journal of Applied Physiology*, **57**, 305–310.

Lindberg, M. (1987). *ARNESE—a PC tool for amplitude analysis of myoelectric signals*. Department of Occupational Medicine and Industrial Ergonomics, Linköping University.

Lindstrand, O. (1988). *Discomfort in the upper extremities with regard to the use of powered screwdrivers in electrical installation work*. Course project, The Swedish National Board of Occupational Safety and Health (in Swedish).

Acknowledgements

This study was supported with funds from the Swedish Council for Building Research (BFR), the Bygghälsan Research Foundation and the University of Linköping.

Chapter 8
Arm suspension balancers in different VDU work conditions

M.A. Tuvesson, K. Ekberg, J.A.E. Eklund, H. Noorlind Brage, P. Odenrick and R. Örtengren

Department of Occupational Health and Industrial Ergonomics, University of Linköping, Linköping, Sweden

Introduction

Visual Display Unit (VDU) work is characterized by static work postures and highly repetitive arm and finger movements. Musculoskeletal disorders are frequent, especially in the neck and shoulder regions. Maintained static muscle contractions are suspected to be a cause of these disorders. Preventive actions are often aimed at reducing the static component and creating increased variation in work. In practice, arm supports or suspensions are sometimes introduced in order to reduce the static component, but the extent to which muscle activity is actually reduced is disputed. A few studies have been performed in order to evaluate the effects of these supports.

The use of arm supports has earlier been shown to decrease shoulder muscle activity in a sitting and resting situation (Andersson and Örtengren, 1974). Lundervold (1951) recommended that arm supports should not be used in typewriting. Sihvonen and Hänninen (1986) found in keyboard work that the activity was lower in the upper part of trapezius when using mobile arm supports compared to working without supports.

EMG-studies in simulated assembly work have shown that arm suspension balancers and arm supports can reduce the neck and shoulder muscle load. The task was highly standardized and consisted of moving a soldering pen slowly over a card within an area of 125 × 125 mm. The reduction was most effective with arm suspension, but the effect was dependent on the sitting posture (Schüldt, 1988). Good acceptance and pain reduction were shown one year after the introduction of arm suspension in an electronics circuit board assembly plant (Harms-Ringdahl and Arborelius, 1987).

In contrast, another study on electronic circuit board assembly showed that shoulder EMG increased when fixed arm supports were used. Nevertheless, the workers were positive about the supports (Sejmer-Andersson and Norin, 1982). Arm support when typewriting or working with a VDU terminal does not necessarily result in a reduction of muscle activity. The use of wrist supports may even increase muscle tension in the upper trapezius during typewriting (Bendix and Jessen, 1986). When operating control sticks, arm support can increase the muscle tension in the trapezius (Lindbeck, 1982).

Hagberg (1988) came to the conclusion that mobile arm supports may be beneficial if the work demands small arm movements. In some situations, arm supports may increase the shoulder muscle tension, especially for larger arm movements. In an effort to compare the effects of arm suspension and wrist support in a data entry task, no clear reduction of the trapezius EMG activity could be seen in either situation (Tuvesson, 1986).

On the whole, it is difficult to make predictions about the efficiency of support/suspension in specific occupations on the basis of previous studies. Arm suspension can be effective as an aid to reducing muscle tension around the neck and shoulder in manual work, but the results are contradictory. Whether arm suspension is beneficial in situations such as VDU work and typewriting has hardly been investigated.

The aim of this study was to assess the influence of an ergonomics aid such as an arm suspension balancer in VDU work, especially data entry. A further aim was to assess the influence of mental stress and to compare this with the influence of the arm suspension balancer. In this chapter, however, only the results from the arm suspension balancer are reported.

Methods

This laboratory study dealt with a VDU task, performed at an adjustable table with an ergonomically satisfactory office chair. The adjustments of the workplace dimensions were chosen by the subjects themselves.

The work task was to use the separate numeric keyboard to enter six-figure numbers, displayed on the VDU screen. Incorrect answers gave a feedback sound. In total, four conditions were used, namely 'stress–arm suspension', 'no stress–arm suspension', 'stress–no arm suspension', and 'no stress–no arm suspension'. The 'no stress' condition meant that the work was to be performed at a pace selected by the subject, and with as many correct answers as possible. The 'stress' condition meant that the correct answer to be given was the displayed number plus 5 for even numbers, and the displayed number plus 6 for uneven numbers. Also, this task was to be performed as quickly and correctly as possible. The influence of stress on the results will only be dealt with very superficially in this paper. In the 'arm suspension' condition, the right lower arm was unloaded by using an arm suspension balancer. The suspension force was individually chosen and adjusted beforehand, and amounted to 5–7 N. The subjects were allowed to become accustomed to the arm support balancer at their ordinary workplaces for 3–4 weeks prior to the experiments. A training session was carried out on a separate occasion before the experiments, and included instruction and familiarization with the task, the experimenters and the measuring situation.

Each experimental session lasted for 30 minutes. There were pauses of at least a quarter of an hour between the sessions. The methods of measurement used were EMG, heart rate and subjective ratings. Automatic records of performance were taken. Surface electrodes on the upper right trapezius muscle were used. Miniature preamplifiers were placed near the electrodes. A normalization of the signal strength to maximum voluntary contraction was made. The resulting muscle utilization ratio (MUR) was expressed in per cent. The EMG equipment was manufactured by TB-Elektronik, Linköping, Sweden. Heart rate was measured via the same equipment and skin electrodes.

Ratings of mood and symptoms were carried out according to Sjöberg *et al.* (1979). The dimensions of activity, relaxation and motivation were rated on a bipolar scale. In addition, pain and discomfort from the neck/shoulders, arms, stomach, eyes and head were rated on a similar scale. The ratings were performed immediately after each experimental situation. Finally, the number of answers and the proportion of correct answers were

Figure 8.1. The experimental work station and a subject using the arm suspension balancer.

collected by the computer on which the task was presented.

In total, nineteen female subjects participated, all professional VDU workers. Their ages varied from 19–56 years, with an average of 36 years. All of them were free from pain or discomfort in the neck and shoulders. Statistical evaluations of EMG, heart rate and performance were made at the 5 per cent level of significance (two-tailed). The paired t-test was used.

Results

The EMG results (in Figure 8.2) showed a tendency towards lower mean values when using an arm suspension balancer, but the values 5.66% and 4.85% MUR respectively did not differ significantly ($p = 0.056$). For the 'static' muscle activity component, defined as the level of the signal amplitude (rms) distribution which was exceeded 90% of the time, there was no difference ($p = 0.63$). No significant difference was found regarding mean heart rate (Figure 8.3) during the sessions ($p = 0.61$).

No significant differences were obtained regarding pain and discomfort from the neck/shoulders, arms, stomach, eyes or head (Figure 8.4). However, there was a tendency towards less discomfort from the arms when using arm suspension balancers, compared with working without balancers ($p = 0.054$). No significant differences were found for the ratings of mood (Figure 8.5), namely activity, relaxation and motivation ($p = 0.39$–0.73). Finally, the mean number of key strokes was 519.5 with arm suspension and 528.0

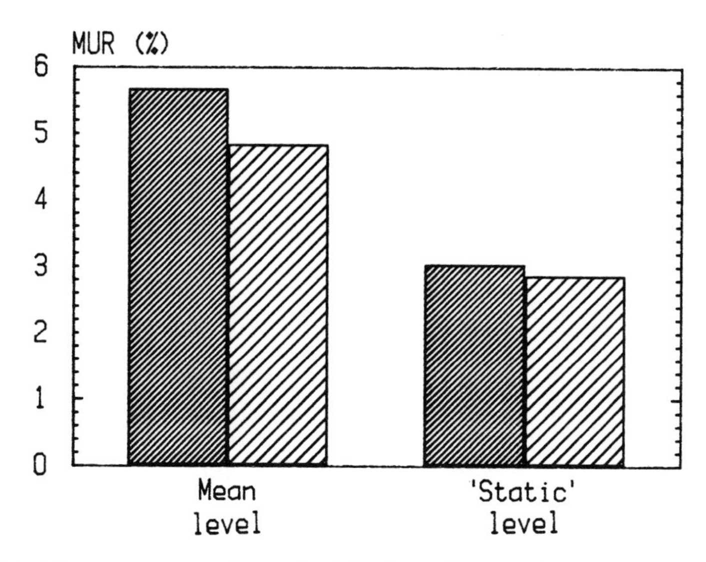

Figure 8.2. The mean MUR and static level for the conditions without arm suspension balancer (left column) and with balancer (right column).

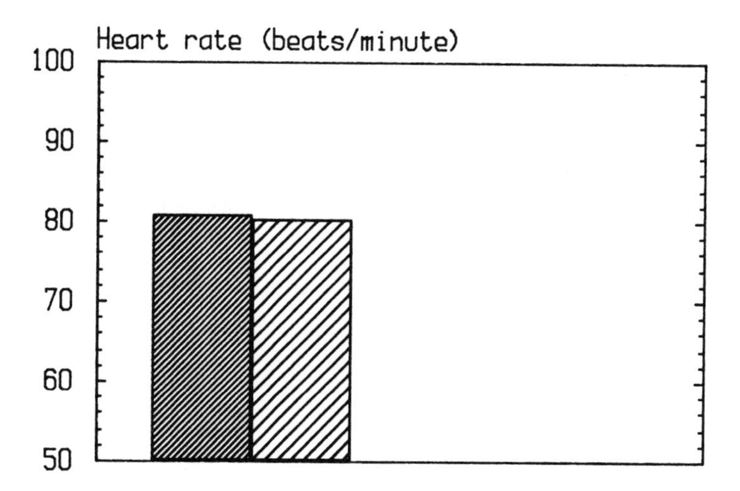

Figure 8.3. Mean heart rate for the conditions without arm suspension balancer (left column) and with balancer (right column).

without. The ratio of correct answers was 97.2% with arm suspension and 97.1% without. These differences were not significant ($p>0.05$).

From sagittal plane pictures, it was estimated that, on average, the moment around the shoulder joint was reduced to almost half of its original value when the arm suspension balancers were used. The 'stress' conditions had little influence on the results. This will be presented elsewhere.

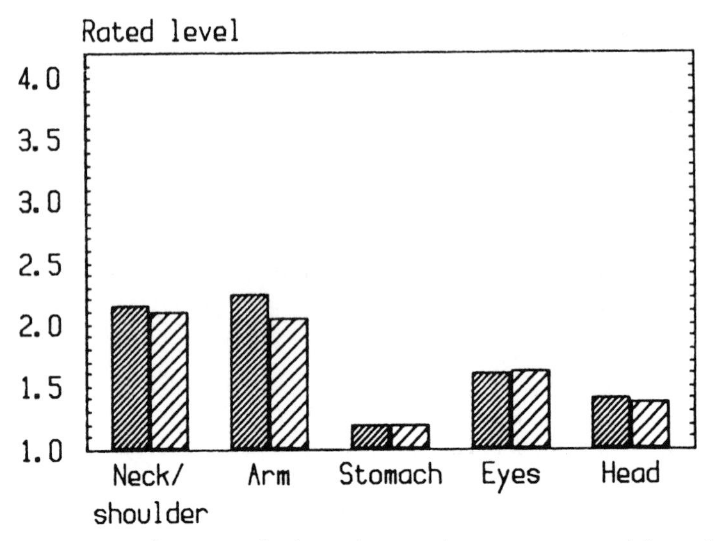

Figure 8.4. Ratings of symptoms for the conditions without arm suspension balancer (left column) and with balancer (right column).

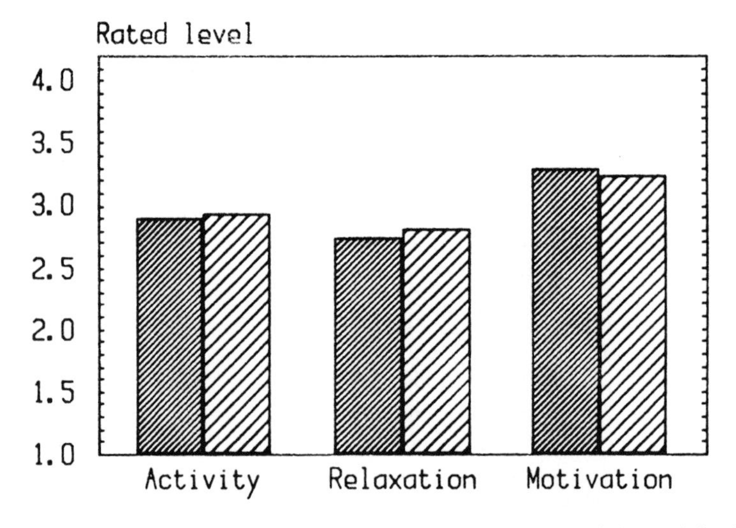

Figure 8.5. Ratings of mood for the conditions without arm suspension balancer (left column) and with balancer (right column).

Discussion

The results show that arm suspension balancers are not a general ergonomic solution or aid in keyboard work. There are, however, large individual differences, which mean that some individuals can obtain benefit from the arm suspension balancers in terms of decreased EMG activity and lower perceived levels of discomfort.

In Sweden, an intensive debate is in progress about the ethics of introducing arm suspension balancers. The critics argue that the use of these aids is unethical, since they in fact tie the person to the workplace and preserve the work posture instead of allowing freedom to move around. Also, this temporary aid contributes to preserving conventional workplace design and organization. Where poor workplace design and organization cause problems, the arm suspenders aim at treating the symptoms, not the cause. Finally, some people consider that the prolonged use of arm suspenders will cause muscular atrophy, although no study supports this. The arguments in defence of arm suspenders are that they have been shown to reduce muscle activity in certain jobs, and will therefore reduce the risk of musculoskeletal problems. There are workers/patients who perceive less pain and discomfort when they use these aids, and therefore they should not be prevented from using them.

There are different opinions in the literature concerning the effects of arm rests. There is no doubt that the work task has considerable influence upon the question of whether arm supports are beneficial or not. The description of the work task is insufficient in most studies concerned with arm support. Two factors are of major concern, namely movements of the arm and the need for stability or precision. The arm support can provide a certain supporting force in the upward direction. The forces in the horizontal direction provide stability, which in some cases is desirable and in other cases not.

From past studies, it seems that when the work task does not require movements of the lower arm or only small rotations around the elbow, there is a good chance that a fixed arm support would be of benefit. However, the use of arm suspension balancers in these situations seems to be perceived as unnecessary and complicated. It can be hypothesized that fixed arm supports are advantageous in this work situation, and in tasks with small and repeated lower arm movements in the horizontal plane, mobile arm supports are advantageous. In tasks which require extensive and repetitive forward reaching not quite in a horizontal plane, arm suspension balancers are the only possible alternative.

The results obtained in this study showed a large individual variation in the EMG values. In combination with individual differences regarding the benefit from the arm suspension balancers, this causes methodological problems in identifying differences, especially when the number of subjects is small. Also, the study indicates effects in the short term, and it is not possible to extrapolate to effects in the long term.

It is obvious that the clear decrease in biomechanical moment around the shoulder joint is not followed by a corresponding decrease in EMG activity. Therefore, the EMG activity is related mainly to shoulder elevation, neck flexion and muscle tensions, i.e., activation also of antagonists due to stress, stabilization or other reasons. The use of a biomechanical model using shoulder moments around the shoulder joint as the main explanatory mechanism for muscle activity is therefore inappropriate in this situation. The more extensive the forward reach in a work situation, the more important the shoulder moment becomes as a source for shoulder EMG activity. When the shoulder moment has become large enough, it will also be the dominating cause of shoulder muscle activity.

Conclusions

The appropriateness of arm supports seems to be determined by the requirements of the task, especially arm movements and the need for stability. Descriptions of the work task are often incomplete in research reports in this field. Arm suspension balancers were shown to have little effect on shoulder muscle activity in keyboard/VDU work. This was especially so for the static component. Arm suspension balancers should therefore not be recommended as a general aid in this type of work. However, some persons experience decreased load on the shoulder muscles and less discomfort when using these aids. The mechanical decrease in shoulder moment when using arm suspension balancers was not reflected by a corresponding decrease in shoulder muscle activity.

References

Andersson, B.G.J. and Örtengren, R. (1974). Lumbar disc pressure and myoelectric back muscle activity during sitting. III. Studies on a wheelchair. *Scandinavian Journal of Rehabilitation Medicine*, **6**, 122–127.

Bendix, T. and Jessen, F. (1986). Wrist support during typing—a controlled, electromyographic study. *Applied Ergonomics*, **17**, 3, 162–168.

Hagberg, M. (1988). *Influence of work environment for occupational disorders in neck and shoulder.* Swedish Work Environment Fund, Stockholm (In Swedish).

Harms-Ringdahl, K. and Arborelius, U.P. (1987). One-year follow-up after introduction of arm suspension at an electronics plant. *Proceedings of the Tenth International Congress of the World Confederation for Physical Therapy, Sydney, Australia.*

Lindbeck, L. (1982). Influence of arm support for shoulder load when operating control sticks. *National Board of Occupational Safety and Health, Report 35* (In Swedish).

Lundervold, A. (1951). Electromyographic investigations of position and manner of working in typewriting. *Acta Physiologica Scandinavica*, **24**, Supplement 84, 115–171.

Schüldt, K. (1988). On neck muscle activity and load reduction in sitting postures. *Scandinavian Journal of Rehabilitation Medicine*, Supplement 19.

Sejmer-Andersson, B. and Norin, K. (1982). *Is the load on neck-shoulder decreased by fixed arm support during assembly work?* Report from Occupational Health Care Center, Ericsson, Stockholm (In Swedish).

Sihvonen, T. and Hänninen, O. (1986). Mobile arm support in keyboard work. Work-related disorders of the neck and upper limbs. *NIVA*, 3–7 November.

Sjöberg, L., Svensson, E. and Persson, L.O. (1979). The measurement of mood. *Scandinavian Journal of Psychology*, **20**, 1–18.

Tuvesson, M. (1986). *Activity in the right trapezius during different ways of unloading the arm.* Occupational Health Care Center, Finspång (In Swedish).

Acknowledgements

The authors would like to acknowledge the devoted work of Maria Carlsson and Mats Lindberg during the analysis.

Chapter 9
Evaluation of hand intensive tasks using biomechanical internal load factors

R. Wells, A. Moore and D. Ranney

Department of Kinesiology, University of Waterloo, Waterloo, Ontario, Canada

Introduction

The wrist, because of its involvement in almost all occupational tasks and its complex and vulnerable structure, is a major site of cumulative trauma disorders. Whilst the main occupational risk factors have been identified, their interaction and safe levels are poorly understood (Armstrong and Silverstein, 1987). Silverstein *et al.* (1986) showed, however, that the two main risk factors, force and repetition, had a multiplicative type of effect. The determination of internal load factors, related logically to the injuries observed and including the effects of the main risk factors of repetition, posture and force over time, is proposed to advance understanding of these conditions.

The relationship between the majority of identified occupational risk factors and one typical disorder, tenosynovitis, is illustrated schematically in Figure 9.1. It is important to note that most risk factors seem to act by modulating the three principal factors. The primacy of repetition, force and posture is emphasised but this also suggests that these three factors, if suitably measured and processed, can account for the effect of many of the other identified risk factors such as vibration or use of gloves. The biomechanical variables logically related to the main types of cumulative trauma disorders of the wrist have been developed from the literature (Moore, 1988) and are summarised in Table 9.1.

Methods

The utility of the load factors introduced above was tested by:

- developing instrumentation and modelling procedures for their determination during occupational tasks
- determining the load factors on a small number of workers performing tasks which had known risks of developing cumulative trauma disorders.

Instrumentation

Five measurements were taken on the workers; surface EMG of the forearm flexor musculature, kinematics of the fingers using a commercial glove transducer, kinematics of the wrist using two potentiometers, video of the worker and force exerted on a staple/

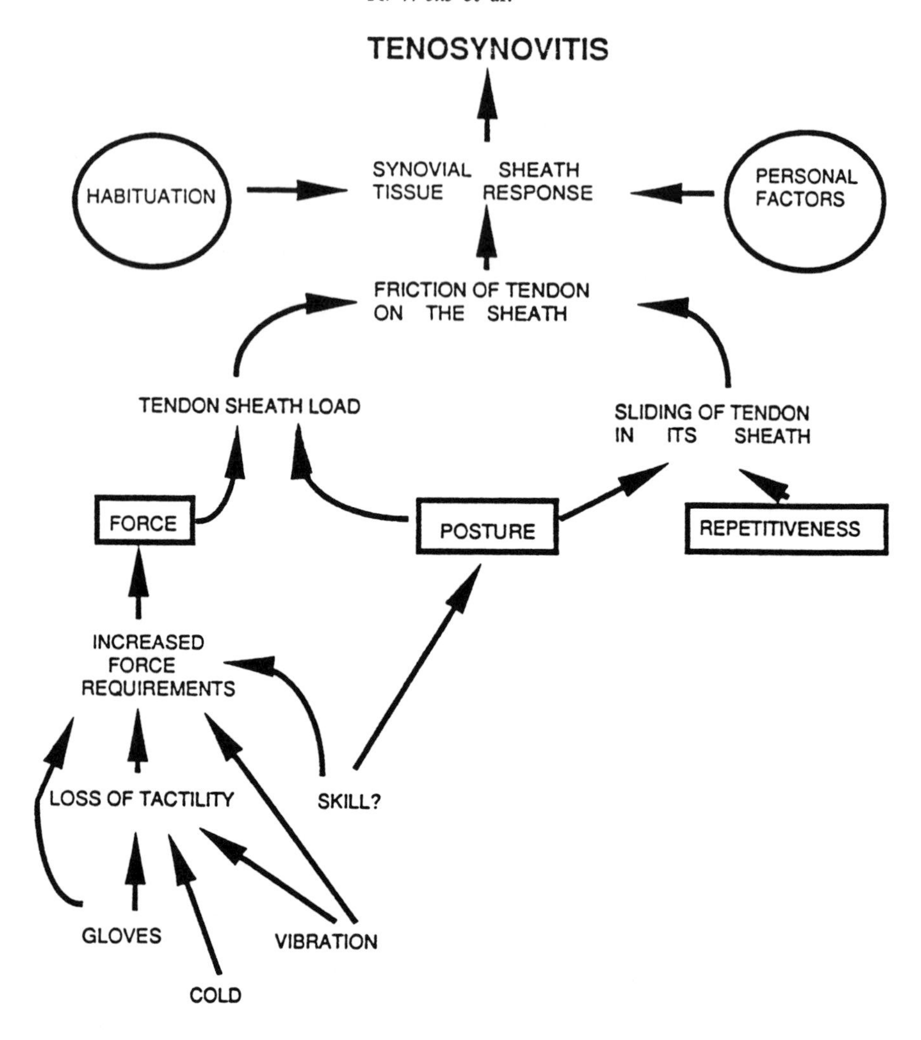

Figure 9.1. Hypothesised relationships between occupational risk factors and the development of tenosynovitis

caulking gun (Wells *et al.*, 1988).

Procedures

The measurements were made on four male workers performing controlled laboratory tasks. A stapling task (high repetition, HIR) and a caulking task (low repetition, LOR) were performed in two force conditions (low, LOF and high, HIF) and in two wrist postures (natural and forced toward flexion). The task classifications were matched to those of Silverstein *et al.* (1986).

These data were used as input to a model of the hand which output the internal load

Table 9.1. The hypothesised relationships between internal load factors and cumulative trauma disorders

CTD	Possible Injury Mechanisms	CTD Risk Factor	Possible Internal Factor
Carpal Tunnel Syndrome	-Force of Tendons on Median Nerve	-Force -Posture -Time	-Peak Normal Pressure -Impulse of Normal Pressure During Flexion
Tenosynovitis	-Force of Tendons on Synovial Sheath	-Force -Posture -Time	-Peak Normal Pressure -Impulse of Normal Pressure During Flexion -Impulse of Normal Pressure During Extension
	-Movement of Tendon with respect to Tendon Sheath	-Repetition -Time	-Tendon Excursion -Peak Excursion Velocity -Average Excursion Velocity
	-Friction Between Tendon and Tendon Sheath	-Force -Repetition -Posture -Time	-Frictional Work Factor -Peak Frictional Power
Tendinitis	-Strain in the Tendons	-Force -Time	-Peak Tendon Axial Force -Impulse of Axial Force
Myalgia	-Muscle Fatigue and Overuse	-Force -Posture -Repetition -Time	-Static Muscle Load (5th Percentile APDF) -Dynamic Muscle Load (50th Percentile APDF) -Peak Muscle Load (95th Percentile APDF)

APDF = Amplitude Probability Distribution Function of EMG

factors chosen for the wrist and extrinsic flexor musculature, Figure 9.2. These included tendon axial force, tendon normal force, tendon frictional work and temporal summations of these variables, as well as the amplitude probability distribution function (APDF) of the forearm electromyogram (EMG).

Results and discussion

Both measured data and calculated output for a high force, high repetition trial in the natural posture are shown in Figure 9.3. Response of the load factors, normalized to the low force/low repetition natural condition, to the different tasks is shown in Table 9.2. In the natural condition subjects had in the order of 30 degrees of wrist extension during the time of peak force. During the forced conditions, the subjects were asked to try to flex their wrists while doing the task. Most of the subjects achieved a relatively neutral wrist at the time of peak force. The difference between the two positions is illustrated in the results (Table 9.2). Load factors such as normal force or frictional work between the tendon and surrounding soft tissues actually decrease in the forced condition because a smaller component of the tendon tension acts in the direction of the surrounding tissues. However, this is contrasted with a significant increase in EMG activity in the forced conditions. This may be the result of two factors. Firstly, in the forced conditions, the muscle is acting at less than optimal length and it has been shown that maximum grasp forces decrease as the wrist goes into flexion (Miller and Wells, 1988). Therefore, a greater percentage of the maximum EMG would be required to achieve comparable force levels.

Secondly, the effort of holding the wrist in flexion could require the activation of the wrist flexor musculature for postural reasons only.

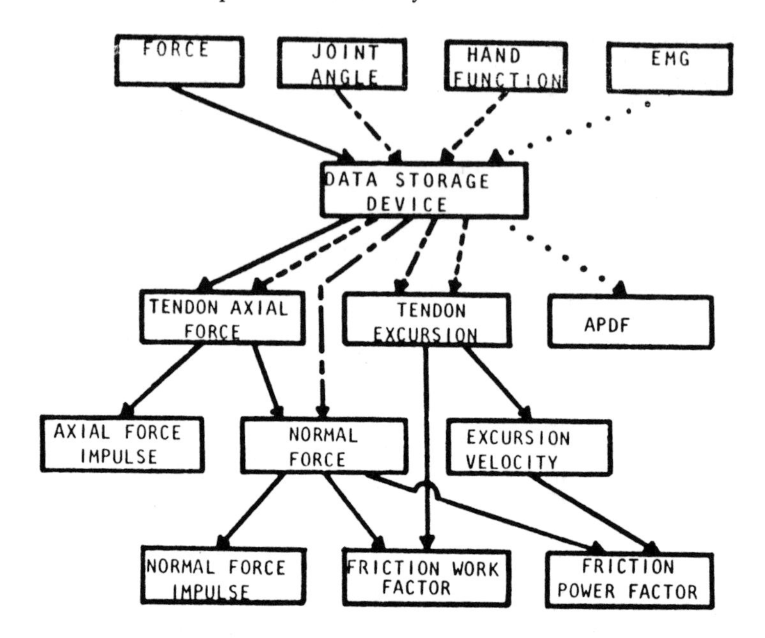

Figure 9.2. Block diagram of the modelling approach and data processing.

Table 9.2. Response of the load factors to four different jobs

Factor	LOF.LOR		LOF.HIR		HIF.LOR		HIF.HIR	
	natural	forced	natural	forced	natural	forced	natural	forced
Peak Axial Force	1.00	.94	1.16	1.07	1.95	1.92	2.23	2.32
Axial Impulse	1.00	.79	1.01	.94	2.42	2.10	3.03	2.80
Excursion	1.00	1.03	3.16	3.46	1.06	1.09	3.37	3.08
Ave. Exc. Velocity	1.00	.91	3.61	3.89	1.08	1.09	3.84	3.46
Peak Exc. Velocity	1.00	1.00	1.31	2.58	1.26	1.31	1.25	.99
Ave. Normal Press.	1.00	.51	.91	.49	2.39	1.01	3.43	.65
Peak Normal Press.	1.00	.50	.92	.65	1.94	1.42	2.51	1.64
Work	1.00	.56	2.75	3.07	1.96	1.15	11.01	1.62
Peak Power	1.00	.62	1.26	1.19	1.78	1.23	5.91	2.00
5th Percentile APDF	/	/	/	/	/	/	/	/
50th Percentile APDF	1.00	1.50	2.00	5.50	4.50	8.00	9.00	13.50
95th Percentile APDF	1.00	1.59	2.05	2.95	1.13	2.37	4.85	6.08

Averaging across postures to match the four categories used by Silverstein *et al.* (1986), the frictional work factor and the 50th percentile APDF of the EMG, averaged across subjects, most closely paralleled the odds ratios of injury found by Silverstein *et al.* (1986): LOF.LOR 1.0; LOF.HIR, 3.3 vs. 3.6; HIF.LOR, 2.1 vs. 4.9 and HIF.HIR, 8.4 vs. 30.3 for the friction work factor and odds ratios of injury respectively. It was also noteworthy that all subjects used much more force than was necessary to accomplish the task. The overgripping ranged from an average of 269% in the HIF condition up to 487% in the LOF conditions.

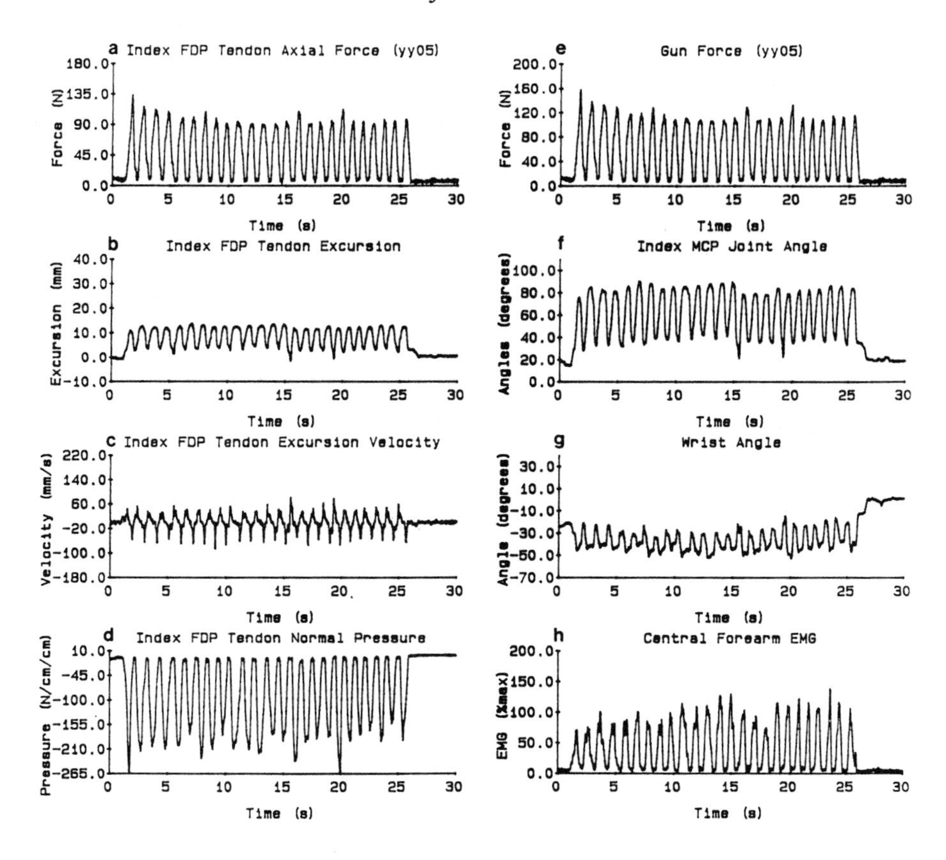

Figure 9.3.

Typical data from a high force/high repetition task (natural posture): (a) Index FDP tendon tension. (b) Index FDP tendon excursion with reference to starting finger position. Positive is into the forearm. (c) Index FDP tendon excursion velocity, positive into forearm. (d) Index FDP tendon normal pressure. Positive against flexor retinaculum, negative against carpal bones. (e) Caulking gun force, measured at gun handle at centre of grip. (f) Index MCP joint angles of flexion. (g) Wrist angle (flexion/extension axis). Positive is flexion. (h) Central forearm EMG. Note: This task was found to be quite demanding by the subject.

The frictional work factor was developed to reflect the transfer of energy from the tendons to the surrounding soft tissues (Moore, 1988). The importance of work in these empirical studies parallels the recent theoretical approach of Tanaka and McGlothlin (1989) who suggest the representation of wrist stress using work, repetitiveness and wrist joint angles as major elements. The authors feel that inclusion of the integrated effect of force and wrist angle (as reflected in the normal impulse factor) would also be important in tasks demanding isometric contractions, such as holding a knife during meat-processing operations. Our choice of tasks in these tests did not allow us to confirm the importance of this factor, however.

The internal load factors described could be potentially useful in ergonomic assessment of jobs for cumulative trauma disorders of the upper limb. This approach allows for quantification of a task in a manner uniquely designed to relate to the mechanical stresses placed on the upper limb of the worker while doing the job. The factors incorporate aspects of the postures, repetition rates, and force required to do the job and combines them at the soft tissue level. Many of the factors use an integrative approach to reflect the importance of time in the development of chronic injuries. The time used could be the hours worked per week or the longest regular period of work without a break.

Conclusions and further work

The measurement system and load factors were useful in accounting for the interaction of the main risk factors for cumulative trauma disorders. The results will also allow for better understanding of the injury-causing mechanisms involved with cumulative trauma disorders. This should permit epidemiological studies to be performed where the hand/wrist exposure variables can be more accurately measured and related to musculoskeletal health outcomes. Such studies in industrial settings are currently underway.

References

Armstrong, T.J. and Silverstein, B.A. (1987). Upper-Extremity Pain in the Workplace—Role of Usage in Causality. In *Clinical Concepts in Regional Musculoskeletal Illness*, (Grune and Stratton), pp. 333–354.

Miller, M. and Wells, R. (1988). The Influence of Wrist Flexion, Wrist Deviation and Forearm Pronation on Pinch and Power Grasp Strength. In *Proceedings of the 5th Biennial Conference and Human Locomotion Symposium of the Canadian Society for Biomechanics* (Ottawa, Canada).

Moore, A.E. (1988). *A System to Predict Internal Load Factors Related to the Development of Cumulative Trauma Disorders of the Carpal Tunnel and Extrinsic Flexor Musculature During Grasping*. Master's Thesis, (University of Waterloo: Waterloo, Canada).

Tanaka, S. and McGlothlin, J.D. (1989). A conceptual model to assess musculoskeletal stress of hand/wrist work. In *Advances in Industrial Ergonomics and Safety I*, edited by A. Mital (London: Taylor & Francis), pp. 419–426.

Silverstein, B.A., Fine, L. and Armstrong, T.J. (1986). Hand Wrist Cumulative Trauma Disorders in Industry. *British Journal of Industrial Medicine*, **43**, 779–784.

Wells, R.P., Moore, A.E. and Ranney, D. (1988). Development of a System for Recording Occupational Hand and Wrist Movement. In *Proceedings of the 21st Annual Conference of the Human Factors Association of Canada*, Edmonton, Canada, Sept 13–16.

Acknowledgements
These studies have been supported by the Ontario Ministry of Labour and the Natural Science and Engineering Research Council of Canada (NSERC).

Chapter 10
Factors affecting health in the workplace

Richard Plette

Human Service, Company for Research and Work Environment Development, 1125 Budapest Diósárok u. 8/b, Hungary

Introduction

The factors which influence health hazards at work places can basically be divided into two classes. One group of factors is generated by the working conditions, the environment and the tasks. The other group is constituted by the social situation, and the individual characteristics and personality of the particular worker.

Work and stress

Any occupational activity or work causes a certain stress. In order to eliminate the harmful effects of work stress we must usually perform some sort of counterbalancing regulatory activity.

People and their environment constitute an inseparable unit; connection between them is necessary, permanent and mutual. The effects realised within this interconnection, independent of their nature, cause changes to the environment as well as in the human organism. While the environmental changes stemming from mutual influence are negligible, at least at the individual level, the changes occurring in the organism may be significant from a biological point of view. To reduce such changes to a minimum needs continuous activity of the adaptive mechanisms.

It must be emphasised, however, that whether it is the individual organism which influences the environment or the environment which produces effects on the organism, in each case it is the organism which suffers because of the stress.

However, to find whether the stress in question (or the change in the organism caused by it) is harmful, neutral or useful, it is not enough just to know about it. What the individual notices is not the stress itself but the changing functions, or a syndrome of these, brought about in the organism by stress. This change of functions is what we call a strain on the organism.

Fatigue

To a certain extent strain is proportional to the load. A greater load, or a stress lasting a longer time, may cause more serious strain. The result of strain is fatigue. The forms in which fatigue manifests itself are usually classified in two groups: peripheral and central fatigue.

Peripheral fatigue causes disorders of the system of nerve tissue connecting the sense organs and muscles with the central nervous system. Central fatigue is for the most part caused by the stress of information processing. The signs of the so-called physiological information load are: disturbances in the sensorimotor faculties, and the disorganisation of motor coordination (i.e. lengthening of the reaction time). The effects of the so-called psychological information load are: disorders in perception, concentration, thinking and/or motivation.

Weariness reduces labour safety. Weariness may be the result either of an individual's inadequate physical condition, or of an overstressed work load, increased work periods, or insufficient resting time.

The role of the worker in labour safety

Working circumstances exert their stressing effects on the working person, whose capabilities to counterbalance these effects are a function of their physical, nervous and psychological characteristics, thus keeping the effects within appropriate limits in order to continue working.

Within the manufacturing or production process human beings are always part of a system. They perform two functions:

1. Equalising the stressing effects of working conditions, to a certain extent by adaptation;
2. Performing controlling functions linked with the work task, by using external sources of power.

Hence, individuals pursue their work within a system of people, machines and working environment in which the risk of accidents grows in proportion to increases in the stressing effects for which they are unable to compensate. With increased physical, neurophysiological and psychological loading they then become unable to control the system adequately.

The causes of stress stemming from working conditions

The stress factors produced by working conditions may emanate from the environment, often qualified by standards. The factors caused by the social conditions of the work are mostly within the control of the management, and the stress factors from the task are, to a certain extent, limited by labour safety regulations, the collective contract of labour and the prescriptions of work norms.

Stress factors caused by the physical environment of work

1. Noise level higher than allowed
2. Insufficient illumination
3. Inadequate climatic conditions
4. Air pollution
5. Harmful vibration
6. Poor technical condition of machines and equipment
7. Insufficient or overcrowded room
8. Standing or tiring working posture
9. Uncomfortable seat or work bench

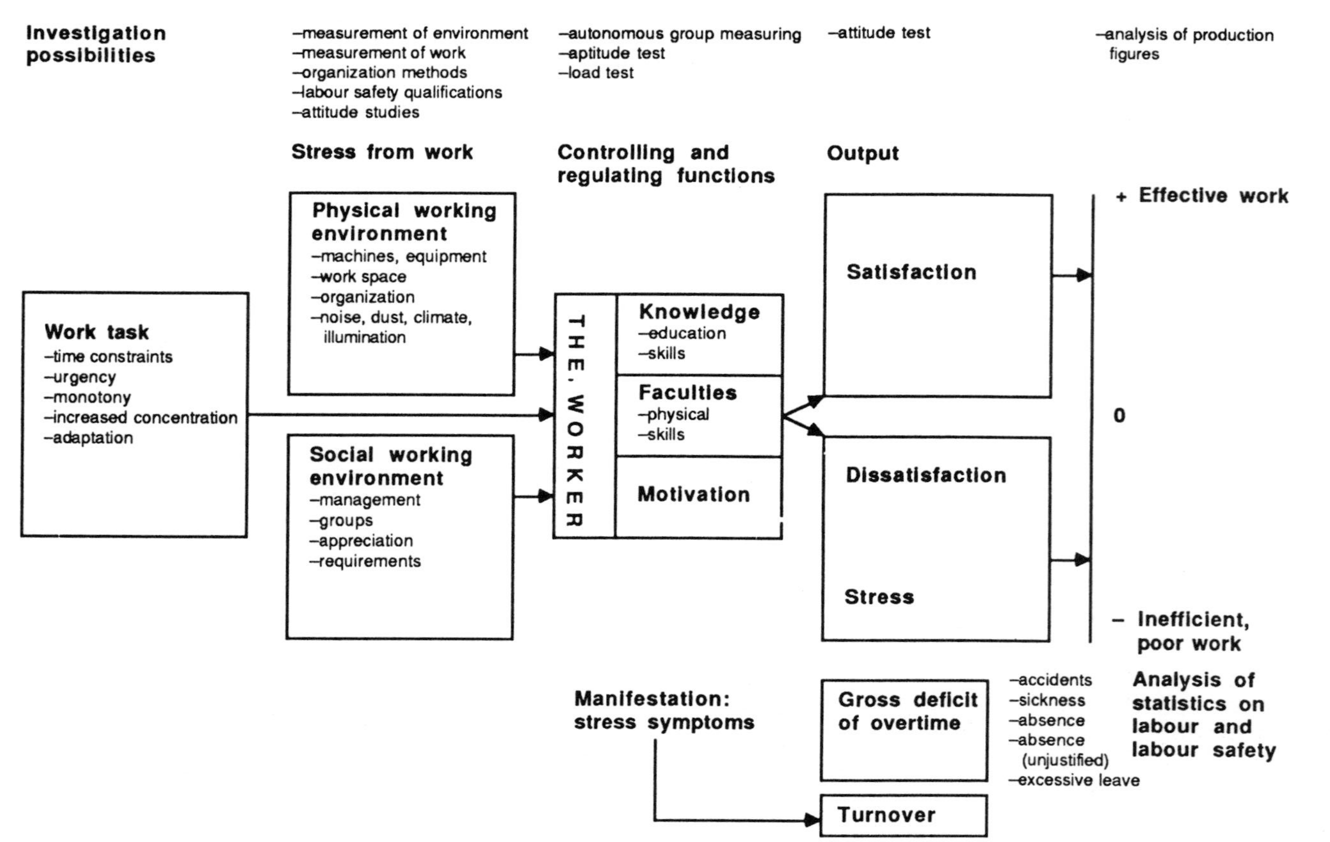

Figure 10.1 Chart of the effects produced on the worker by the production process

Stress factors caused by social environment

1 Disorganisation, unequal work loads
2 Hasty, nervous atmosphere
3 Little appreciation of regulations
4 Tensions among workmates
5 Tensions caused by management: (i) inadequate professional skills; (ii) improper guiding style; (iii) lack of respect; (iv) favouritism
6 Frequent contradictory instructions
7 Low wages or salaries compared with the job demands
8 Lack of confidence by management or too rigorous inspection systems

Stress factors stemming from the work task

1 Regular overtime
2 More than one shift
3 Increased pace of work
4 Constraints of time, pacing of work dictated by the machine
5 Unrealistic targets
6 Heavy physical work
7 Short term of production, consequent urgency
8 Monotony
9 Increased need for attention
10 Responsibility for own physical health
11 Responsibility for others' physical health
12 Frequent dangerous situations
13 Frequent adaptations to others
14 Frequent breaks in the work process or resettings, etc.

The personal factors of stress

The effect of increased hazard from loads at the workplace shows some correlation with certain personal factors. These should be handled carefully, since their existence only means a risk if they accumulate and if they become active as a result of extensive stress conditions of work. Such personal factors are:

1 *Risk factors caused by sociological state of the individual:* (i) unmarried; (ii) age below 25 years; (iii) male sex
2 *Risk factors of social situation and way of living:* (i) disorderly family life; (ii) uncertain or poor housing situation; (iii) family budget problems; (iv) high alcohol consumption
3 *Risk factors linked to faculties and capabilities*; (i) deficiencies in sense organs; (ii) slow reaction time; (iii) poor concentration; (iv) difficulties in perception; (v) inflexible ways of thinking
4 *Risk factors of personality and motivation*; (i) extreme extroversion; (ii) neuroticism; (iii) exaggerated risk taking; (iv) impulsive behaviour, searching for adventures; (v) a behaviour strategy for avoiding failures; (vi) negative attitude to labour safety; (vii) over-sensitivity to stress factors

Response of workers to stress from the work environment

Workers can react to the stressing effects of working conditions in different ways. If the factors of the work conditions remain within tolerable limits, the worker is able to equalise their stressing effect and perform balanced, efficient and safe work. If, however, the stressing effects exceed limits of the tolerable range, i.e. beyond the adaptation capacity of the individual, then the connection between the person and the operating conditions will be destroyed and the individual will react by exhibiting a state of stress as a means of protecting him or herself and, as a result, stress symptoms will appear. This stress is a specific state of the psyche in which a person feels overstrained and powerless, unable to cope with the load.

The symptoms of stress

Characteristics of the stressed state are:

- permanent fatigue, nervousness and irritability, anorexia and insomnia
- aggressive behaviour which after a certain time may change into lethargy and indifference
- the so-called withdrawal instinct manifesting itself in harmful habits, such as alcohol, smoking etc.
- irregularity of actions, irritability and hastiness, as in the signs of panic
- various disorders indicating anxiety, such as stuttering, inhibitions etc.
- certain psycho-physical symptoms, such as troubles of attention, decreasing intensity, rigid behaviour, mental deficiencies, amnesia, inability to concentrate etc.
- weak-spiritedness, apathy, indifference
- resistance to being convinced, or to changing opinions

The results of stress

- decreased performance both in quantity and quality
- increased absence from work
- increased number of accidents and occupational diseases
- development of industrial fatigue, neuroses, or absence through sickness
- increased labour turnover

Study of health causes and effects at work

Investigations should be conducted urgently into the causes of health hazards at work. The first task is to produce a weighted evaluation of the stress factors of the working conditions which result from the physical environment, social environment and the tasks of work. This means identifying the degrees of strain or fatigue which are caused by particular stressing conditions of work, or in other words, the extent to which they contribute to the development of strain. If this could be achieved, we should be able to evaluate each stress factor in actual cases in particular organisations. This would allow identification of the factors which influence health hazards at the actual work places (see Figure 10.1).

The next step would be to initiate a detailed analysis of the nature of the factors which are actually present. These could be investigations into technology, organization, working conditions, or social psychology. Parallel with such a series of studies, the personal characteristics of the workers would also have to be examined.

Part IV

Evaluating the workplace

The discussion of workplace evaluation has been divided into two halves, the first reviewing the requirements and limitations of the techniques which are currently available for ergonomic workplace analyses, while the papers in the second half describe their application.

Many of the questions in workplace design relate to layout of the equipment and workspace, and CAD (computer-aided design) systems are likely to become widely used for this purpose. They also provide both an opportunity and a challenge for ergonomists to gain the support of engineers and designers in order to make the changes which are needed in the workplace. We need therefore to consider how well ergonomic information and style of analysis can be matched to other parts of the design process, and it is likely that the method of presentation will become an important issue in the design of the modelling systems themselves. For example, engineers will understand the biomechanical concepts quite readily, but may be less familiar with physiological limitations, so that this in turn is likely to involve concern issues of training in the use of CAD systems and perhaps ergonomic education more generally.

Örtengren and his colleagues looked at several of these questions in a simulation of an integrated system incorporating workspace and biomechanical evaluations. Their experience with designers was very positive and leads to recommendations for facilities which should be incorporated to make this a valuable tool in the production planning process. Kuusisto identifies further features which are important in computer-aided ergonomic design, and discusses the potential of expert systems to assist in workplace investigations.

Randle, Nicholson, Buckle and Stubbs present a critical review of materials handling guidelines, and once again question whether our ergonomic models and databases are appropriate or whether they need so much interpretation that they require expert ergonomic users. Their conclusions show that even ergonomists have difficulties in applying the models to industrial situations, and that we still need a much better understanding of the interacting influences of biomechanical, physiological and psychophysical factors involved.

Turning from workplace design to workplace redesign, Keyserling and Armstrong describe a structured approach to job analysis which has helped in identifying risk factors which might result in overexertion injuries, and discuss how this can be combined with a participative approach applied by a team with a core membership of the operator, the supervisor and an ergonomist.

Applications of CAD systems in the design of a variety of workplaces are described in the two papers by Launis and Lehtelä on the 2-D ergoSHAPE microcomputer-based system and by Case, Bonney, Porter and Freer on the 3-D SAMMIE CAD system. Their work with design teams also provides feedback on future developments which are needed within ergonomic CAD packages.

Chapter 11
Ergonomic CAD systems in work design: an investigation of requirements and limitations in a practical application

Roland Örtengren, Margareta Liew and Mats Torgnysson
Department of Industrial Ergonomics, University of Linköping, S-581 83 Linköping, Sweden
Bo Glimskär
BELAB Company, Björnnäsvägen 21, S-113 47 Stockholm, Sweden
and Gunilla Myrnerts and Rolf Dandanell
SAAB-SCANIA AB, Department of Occupational Safety and Health, S-581 88 Linköping, Sweden

Introduction

In the mechanical engineering industry CAD systems are important tools used by engineers to speed up the design process and to simplify the production of drawings. Large manufacturing companies, and also many small ones, have changed to CAD almost entirely. The organisation is such that the designers work at CAD terminals connected to a large host computer. The drawings produced are stored in the memory of the host computer, and thus they can be shared by several designers working on the same or adjacent details of the whole design.

Although this may cause problems in itself concerning responsibility for changes, authorisation, data base management, safety and secrecy and so on, no part needs to be redrawn once it has been completed and drawings of standard parts can be called up from a common library. Since in many cases new designs are only modifications of old products, new drawings can easily be produced by taking copies of the old ones and introducing the changes. At the start of this development, the CAD systems were used as drawing tools and the drawings still defined the product. Now, with CAD systems having full three-dimensional capability, a design can be defined by what is stored in the computer, which is a great advantage for complicated geometries such as the curvature of surfaces.

Part of engineering design is to check material strength and safety margins of a particular geometry. Strength calculations are made using so-called finite element methods (FEM). To expedite this procedure, programs for FEM calculations use the same geometrical representations as the CAD system. CAD systems therefore have FEM software as subpackages so that the calculations and the drawings can be made in the same computer.

Through integration with manufacturing (CIM), it is possible to transform descriptions of surfaces into programs controlling cutting sequences for milling machines. Great efforts have been devoted to the development of CIM technology because it creates the potential for a rationalisation of production engineering and production planning which is considered to be very large. This potential for rationalisation exists not only in relation to manufacturing but also to assembly. Here the benefit is that the production engineering work can start early, as soon as tentative designs are created.

CAD systems have also been used by ergonomists for workplace design. By introducing an anthropometrically correct model of a human operator, ergonomic consequences of product design as well as working environments can be considered. Many computerised man models have been developed, but only few have been introduced into commercially available CAD systems (Hickey *et al.*, 1985). One example is the SAMMIE (System for Aiding Man-Machine Interaction Evaluation) system (1987) for workplace and product design.

Man models have also been created for applications in computer graphics. Here the purpose has been to produce images and films of products and surroundings, mostly for commercial use but also for scientific purposes. For example, car manufacturers are using computer graphics to create images in their work on new model designs which they expect will lead to a reduction of the number of mock-ups and clay models needed. A special challenge has been in animation, i.e. to model human and animal movements using articulated figures (Badler, 1987). CAD systems do not normally have facilities for animation, nor is there any possibility for surface rendering other than solid modelling.

An ergonomic analysis of a workplace can be based on a three dimensional visualisation of the working environment, introducing properly sized human operators. The evaluation can be based on judgements made by an ergonomist who visually checks the postures and reach distances. The SAMMIE system allows this and it can also signal when 'comfort values' of joint angles are exceeded.

No system yet available, however, permits the calculation of physical loads by means of a biomechanical model, although geometrical descriptions of posture can be obtained from the system. The aim of this study has been to explore these possibilities and to evaluate the prospects for their future use in production planning. To that purpose such a link was simulated and an ergonomic evaluation of an existing workplace was made using an ergonomic CAD system as well as a traditional approach. In addition, design engineers and occupational safety and health personnel were interviewed about their views on computerised ergonomic planning and evaluation.

Methods and procedures

Production work

The work selected for study was assembling part of the nose portion (called lower structure) of the SAAB 340 civil aircraft in the SAAB-SCANIA manufacturing plant in Linköping. The lower structure consists of the nose cone, the main instrument panel, the cockpit floor and the structure below that. It is built from pre-shaped parts and sheets of aluminium which are riveted together. The structure is built in a jig in order to define its measurements very precisely.

Riveting is a precision task, especially demanding in the aircraft industry with its high safety standards. The procedure is lengthy since it is done in several steps. First, the parts are temporarily joined together and holes for the rivets are drilled. The parts are then separated and cleaned, sealing compound is applied and they are joined together again. The riveting is done by means of a pneumatically driven hammer operated by one worker while another applies a restraining force to the other side. This is done mostly from below. In some riveting work, the same worker does both tasks, operating the hammer and holding the resistance. During the work the quality of the result is checked several times.

The jig for the lower structure had a main floor about 2 m above the shop floor where most of the work was done. The cockpit floor was situated 30 cm above the main floor and, to facilitate work on the sides of the structure, there were lower floors covered by removable hatches in the main floor.

In the manufacturing hall there were two jigs where lower structures were assembled and another workstation for supplementary work. It takes two weeks to build one structure and one week to finish the supplementary work. Altogether six persons worked on the lower structure, two at each workstation. Their mean age was 23 years (20–25 years) and their mean time of employment was 3 years. The payment was a combination of piece and quality rates shared between the two members of each working group.

Five work tasks, which could be distinguished from each other by posture, position or task, were selected for ergonomic analysis. They were all considered to be particularly strenuous.

Jig design

Jig design is part of the production planning process. It starts as soon as engineering designers have decided on the general shape of an aircraft. The jig designers work in parallel with the engineering designers and they have daily informal contacts. The goal is that the design of a jig should be finished at the same time as the product design.

The jig designer is constrained by the shape and construction of the aircraft, but otherwise many design parameters can be varied and he can choose the jig layout he believes to be optimum. He can determine the size of the part which will be built in a single jig, the operator's working posture either standing or lying, the height of the jig and the number of people working simultaneously. Thus there are many parameters that must be determined to reach an optimal solution considering consequences for production time, production cost, quality and ergonomic conditions. Using his experience the jig designer presents drawings of a few alternative designs of which the best is then selected at a production planning meeting.

The jig designers have a good knowledge of production and the problems that arise. Although there are frequent contacts between the designers of the aircraft and the jig designers, the jig designers are not likely to influence the design of the aircraft itself when they foresee production problems—and are even less likely to influence ergonomic problems. Only if alternative solutions come up during the design work can the working environment be taken into account. This is, moreover subject to the additional constraint that the designers must first be aware of the production problems.

When a drawing of a jig is made, it can be difficult to see if the jig will work in practice. As an aid in judging how, if at all, a work task can be carried out, the designers use manikins which they place on the drawing. The manikins are available in different scales and anthropometric sizes. No hint of physical load is obtained, and the illustration is only two dimensional. In particularly difficult cases, a mock-up model of the jig is made, sometimes in full scale. The jig can then be tested using real people, otherwise 3-D manikins are used. The jig designers make the scale models to convince themselves, the management and the workers that the working conditions will be good. Although the models are expensive to make, they are generally considered necessary to illustrate how a workplace will look in reality

CAD simulation of work and ergonomic evaluation.

For the ergonomic evaluation in the present study, the SAMMIE system (Price Computer Inc., 1984) was used. The system is a design tool which has been developed to allow 3-D computer modelling of products, buildings and people. The system permits evaluations of postural comfort, and the assessment of clearances, reach and vision to be conducted on the earliest designs. The man model in the system is a 3-D representation of the human body with articulations at all the major body joints. Limits to joints can be specified and the dimensions and body shape of the man model can be varied to reflect the range of sizes and shapes in the relevant national or occupational population.

To build a geometrical model in the SAMMIE system, engineering drawings of the lower structure and the jig were obtained. The original drawings were very detailed and therefore they had to be simplified and several had to be combined in building the model. Ideally the design data of the structure should be transferred electronically from the CAD system where it was constructed, but as the jig in this study was designed on paper, only paper drawings were used. This was an advantage here as it became easy to simplify the model (Figure 11.1).

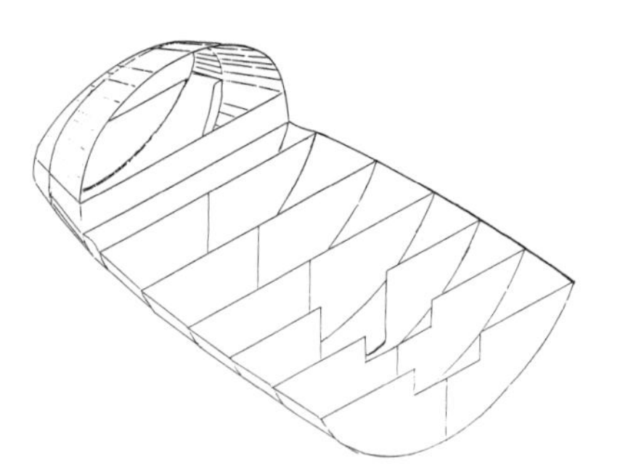

Figure 11.1. Simplified drawing of the lower structure of the SAAB 340 aircraft for modelling in the SAMMIE system.

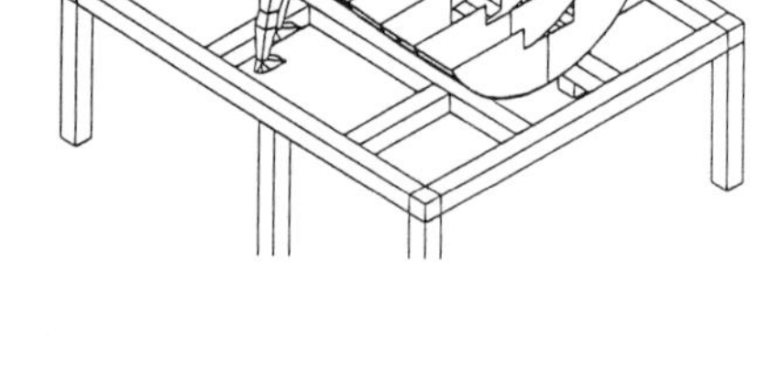

Figure 11.2. View from the side of an assembly worker riveting aluminium sheet to the bottom of the lower structure as modelled in the SAMMIE system.

A special task scenario was created for each of the five work tasks selected for evaluation (Figure 11.2). In each an operator model was introduced and specified by the anthropometric data of the people actually doing the job. A biomechanical evaluation of the work postures was made be feeding postural data from the model to a two-dimensional strength prediction program (Chaffin and Andersson, 1984) by means of which loads at major joints were calculated (Figure 11.3). The work was also filmed on video and the working postures seen on the film were compared with those modelled in the CAD system. The operators were asked to complete standard ergonomic questionnaires and also asked about their experiences of comfort/discomfort in different body parts during work, using a method based on that of Corlett and Bishop (1976).

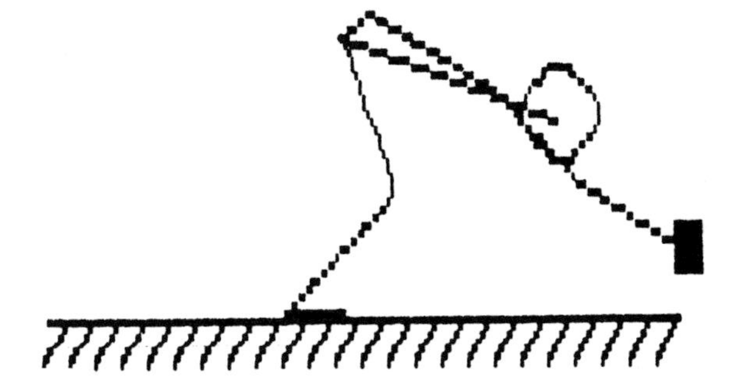

Figure 11.3. Simplified drawing of the operator in Figure 11.2 used in the strength prediction program for biomechanical calculations of body loadings.

In addition, the whole evaluation procedure was demonstrated to the production engineers and the designers who were responsible for the lower structure jig (the engineers), as well as to staff members of the occupational safety and health department. The function and facilities of a tentative system for ergonomic evaluation were explained to them and their opinions were sought about the usefulness of the system and its requirements.

Results

The results of this study which have reference to the procedures used by the jig designers and the production engineers have been described above. The results of the ergonomic evaluation show that it is possible to perform a biomechanical analysis using video records of work postures as well as work postures modelled in a CAD system. The analyses yielded similar results when the work postures to be evaluated were similar. As the analysis is based on still pictures of working postures, it is important that relevant postures are selected by the evaluator, which requires a certain amount of ergonomic experience.

The subjective evaluations made by the workers showed that the work situations chosen were particularly strenuous and that no work situation had been missed. The body parts that were subject to high discomfort were also the ones for which the biomechanical analysis indicated a high ratio between utilised and available body strength. Thus it was possible to pick out postures which were limiting performance from that point of view.

In the analysis of the subjective assessment of discomfort and pain, measurements were repeated and the development of the sensations could be followed during a working day. The subjects had different tasks, but several of them experienced discomfort that increased to pain over the day. Such results could not be obtained from the biomechanical analyses, which gave the same load irrespective of the time of day.

An ergonomic analysis of workplaces by means of a CAD system with a human operator model was found to be feasible and could be made in the early planning stages. However, it is obvious that the analysis must be supported by evaluation procedures that are easy to use. Some requirements for the program specification can be formulated.

The potential users, the jig designers, were of different opinions about CAD systems with capability for ergonomic evaluation. Some were even doubtful about the usefulness of CAD systems at all in their work, as they considered their present paper-based routines to be adequate both for design of the jigs and for evaluation of certain job conditions. Also their opinions varied about the value of ergonomic evaluation of postural stress; the interest of one designer, for example, being restricted reach distances and dimensions of openings or passages. Several engineers thought that they should not do any ergonomic evaluation because they had neither the time nor the competence.

The occupational safety and health staff, on the other hand, who daily see patients suffering from bad working conditions, were convinced that all work tasks should be subject to ergonomic evaluation, but they also thought that they should not do the evaluation since they did not have the time.

Discussion

The results of the ergonomic evaluations made in this study show that it is possible to reach the same results using an ergonomic CAD system as with a traditional biomechanical approach. Thus it is important that an ergonomic CAD system has facilities for workload calculations by means of a biomechanical model. As the long term effects of the workload were found to be important, it should be possible to consider also the time a task takes and the relationship between work and rest periods, for example using the computerised method suggested by Glimskär *et al.* (1987).

The jig designers mentioned a number of requirements that a CAD system for ergonomic evaluation should fulfil based on their present experience of ordinary CAD systems. The ergonomic facilities should be easy to use with simple menus and commands. The system should have an effective way of controlling the position of the man model, which otherwise can be both tedious and complicated if the number of articulated links is large. The system should signal when a strenuous or fatiguing posture is adopted. Some designers mentioned that they wanted an expert system to support them in the ergonomic evaluation.

The experiences of the use of the SAMMIE system gave similar requirements with a few additional ones the most important of which was that the system should be sufficiently fast that redrawing the screen to show new postures takes only a couple of seconds. There should be a tabular display to show numerical coordinates of the position of the man model. Another important feature would be automatic signalling when the man model made contact with an object in the vicinity. Otherwise it would be very difficult to adjust the model to stand properly on the floor or to sit on a chair, or to know when the surface of an object was touched by the hand.

The engineers require that the ergonomic evaluation can be made in the same CAD system that they normally use. If a separate system were used, the transfer of objects and geometrical relations should be made very efficiently with adequate filtering of unnecessary details. All agreed that CAD systems with enhanced graphics capabilities and animation will lead to both better and more functional products as well as better conditions for the production workers, but that systems currently available need to be developed further and to be made simpler to use.

The development of CAD systems with a man model for ergonomic evaluations has been slow, although work is going on in many research laboratories. One reason for this is the low demand from industry; another is that ergonomic evaluation is complicated and that there are no established procedures into which ergonomic CAD systems would fit. The systems available so far have been applied in product design and not in workplace and equipment planning, and mostly by expert ergonomists in consulting work. None of the systems is compatible with major CAD program systems, which is considered to be a serious drawback as time must be devoted to constructing models that are already available in electronic form. Exchange of information between CAD systems has become increasingly important, however, and the developments in the area are very fast at present. Another problem with ergonomic CAD systems is that they have not reached the same level of functionality and efficiency as ordinary CAD programs, which makes their acceptance by potential industrial users even more difficult.

The demands from industry have increased in recent years because of the drive to reduce lead time at all stages of product development and production planning. Thus there is a need to increase communication between engineers and to reduce the number of mock-ups and prototypes being made for testing purposes. Working conditions have also received greater concern from the management who need to use good working conditions as an attraction in the competitive labour market.

It often happens that the poor working conditions that are seen in manufacturing and assembly operations have been caused by decisions taken at early stages of product development. During the product development process, decisions are taken sequentially and the further the process has gone, the more difficult it becomes to make changes. Sometimes problems arise because the engineers are not aware of which factors are likely to give problems during production, sometimes it is because the decisions have gone so far along the line that it is no longer possible to change anything for organisational reasons. When a problem is left to a later stage of the planning process, however, the cost of its remedy increases dramatically.

No doubt the possibility of making changes in today's organisation would increase if the consequences of bad design could be foreseen earlier. This will be essential if lead time is to be reduced (i.e. the time it takes to put a new design into production) and there is a strong demand for this. A study of the planning procedure shows that to overcome some of the difficulties it is necessary to delay more decisions to a later stage and to increase the information exchange between the different stages. Information must also be available earlier, perhaps already at the sketch and concept stage.

The problem with insufficient awareness could be improved if more people (with the right competence) communicated with each other during the planning process. Within a department, the exchange of information between people is probably sufficient, but between departments this is difficult to accomplish. The possibility of using different organisation to simplify and enhance information exchange was tested by Ford some years ago. The planning and development work was done by a program team in which representatives from planning, design, engineering and manufacturing and even assembly workers participated (Cortes-Comerer, 1987).

An increased exchange of information between departments in the present organisation could be accomplished by, for example, a shared CAD system with enhanced capabilities in that respect, where product geometries and present status of a design project were made available to everybody involved. Here an ergonomist could easily be included to be responsible for the human factors side of job design. Increased use of computers for information exchange and greater attention to human resources management are current trends in industrial development (De Vries, 1988).

Systems for computer graphics that can be connected to existing CAD systems will have an important role here as they are made available to the engineers. The graphics system can be used for modelling and animation, thus providing an effective tool for visualisation of structure and function of a product as well as of the working environment in production. The importance of this is to facilitate understanding at meetings where decisions are to be taken about design features.

There are several factors which are important in implementing procedures for ergonomic evaluation of work tasks at the planning stage. First, there must be an incentive to do the early evaluations and this will not come from the lower levels of the organisation. The company must adopt a strategy for the working environment stating that, as one step, evaluation of ergonomic consequences must be made before all planned work tasks are accepted. At present the engineers are not motivated to evaluate a workplace, apart from checking that the geometrical and physical conditions make it possible to carry out the job at all. Physiological work load and postural strain are not taken into account.

Second, there is the question of who is going to do the evaluation. The engineers are in the right place in the organisation, have the technical knowledge about production methods and can identify alternative solutions when needed, but they do not have the ergonomics knowledge required to perform the evaluations without additional support. The personnel from the occupational safety and health department have the proper ergonomics knowledge, but their technical competence is not sufficiently high and they belong to a different department. For organisational reasons, the engineers should be required to do the ergonomic evaluations but, in support, they must be provided with more and longer courses in ergonomics than the two days that some of the jig designers in this study had.

Conclusions

A CAD system is a necessary tool to enable ergonomic evaluation during production planning, and needs an anthropometrically correct and adjustable man model with adequate software for calculations and control. Such a system would make it possible to identify and eliminate strenuous work tasks at the early stages of production design. The ergonomic analysis should be included in the ordinary CAD system, or the system used should be able to communicate with the ordinary CAD system. It is suggested that the engineers working with production planning should do the evaluations, and that they should be given a better ergonomics education. The ergonomic CAD system should be provided with built-in expert knowledge to advise and support the engineers in the evaluation. The goal should be that the engineers should be able to handle and eliminate ergonomic problems in 90 percent of the cases. For the more complicated cases it should be possible to call in an ergonomics expert or group of experts who should also have the responsibility for methods and systems development.

References

Badler, N.I. (Guest Editor) (1987). Articulated figure animation. *IEEE Computer Graphics and Applications*, **7(6)**, 10–11.

Chaffin, D.B. and Andersson, G.B.J. (1984). *Occupational Biomechanics*. (New York, NY: John Wiley & Sons).

Corlett, E.N. and Bishop, R.P. (1976). A technique for assessing postural discomfort. *Ergonomics*, **19**, 175–182.

Cortes-Comerer, N. (1987). Organizing the design team. *IEEE Spectrum*, **24(5)**, 41–46.

De Vries, M. (1988). The industrial renaissance: Technological trends and educational implications. *Newsletter of IEEE Council on Robotics and Automation*, **2(2)**, 5–9.

Glimskär, B., Höglund, P.-E. and Örtengren, R. (1987). *Ergo-Index. A description of ergonomic effects*. Report, TRITA-BEL-0036. LiTH-IERG-R-9, (Stockholm: Royal

Institute of Technology) (In Swedish).
Hickey, D.T., Pierrynowski, M.R. and Rothwell, P.L. (1985). *Man-modelling CAD programs for workspace evaluations*. Report, Defence and Civil Institute of Environmental Medicine, Downsview, Ontario, Canada.
Prime Computer Inc. (1984). *SAMMIE. Human factors 3-D design system*. Product Bulletin available from Price Computer Inc., Natick, MA, USA.
SAMMIE CAD Ltd (1987). *The SAMMIE system*. Information booklet published by SAMMIE CAD Ltd, Loughborough, UK.

Acknowledgements

Financial support for this project was provided by the Swedish Work Environment Fund, SAAB-SCANIA AB and the University of Linköping.

Chapter 12
Limitations in the application of materials handling guidelines

I.P.M. Randle, A.S. Nicholson, P.W. Buckle and D.A. Stubbs

Ergonomics Research Unit, Robens Institute, University of Surrey, Guildford, Surrey GU2 5XH, U.K.

Introduction

Manual materials handling has been a focus of attention amongst the ergonomics and biomechanics research communities for more than twenty years. Much effort has been directed towards measuring and modelling the handling capabilities and capacities of the working male and, to a lesser extent, female populations. This has resulted in a number of models or data sets of manual materials handling activities being developed with the aim of helping to reduce health related disorders, usually of the musculoskeletal system, and/or to improve the efficiency of the work system. These have been based on biomechanical, physiological, or psychophysical criteria, or a combination of these. Models or data sets have been produced of lifting, lowering, pushing, pulling or carrying tasks, with some consideration being given to combinations of these variables. These have gained a wide audience and general acceptance amongst practitioners, health and safety personnel and the legal profession.

As a result of 'hands-on' experience of these various guidelines, the authors have become aware of their limitations and that their indiscriminate use could have serious implications for the effectiveness of workplace and task design. This chapter investigates these limitations by first examining the manual handling system and the extent to which the guidelines incorporate the system components. Secondly, it refers to a study carried out to compare qualitatively the most commonly used guidelines. Finally, it considers three industrial case studies where guidelines have been used to recommend acceptable manual handling forces for a particular work task.

Guidelines and their uses

In examining existing manual handling guidelines, it is apparent that many have been produced as a result of funding bodies providing sponsorship to address a particular set of circumstances pertinent to their own requirements. Others have been the result of laboratory studies addressing particular health and safety aspects of manual materials handling such as lifting, lowering, carrying, pushing and pulling (see Table 12.1). As a result, whilst these models or data sets may be scientifically valid, the guidelines produced

Table 12.1. Summary of manual handling models and guidelines

Task(s)	Reference	Output	Required inputs	Limitations/uses
Lift, lower, carry, push/pull	Garg *et al.* (1978)	Metabolic rate	W,L,G,Ho, Hd,S,X,F	Stoop, squat and arm lift, static pushing force, bimanual carry
Lift, lower	Asfour (1980)	Oxygen consumption	F,L,H, box size	Lift from floor + table, + twist, lower to floor level
Lift	Mital *et al.* (1984)	Oxygen consumption	Sex, strength, F, box size, duration, lift capability	Lift all heights
Lift	Karwowski and Ayoub (1984)	Oxygen consumption	Lift capability, W, F, age	Lift from floor to 0.76m
Lift	NIOSH (1981)	Load weight at action limit & max permissible limit	H,Ho,D, Fmax, F	Bimanual sagittal plane lifting
Lift, push/pull	Materials Handling Unit (1980)	Acceptable weight	X, Y & Z coordinates of hand, posture	Uni/bimanual, based on intra-abdominal pressure and arm strength
Lift, lower, carry, push/pull	Snook (1978)	Acceptable weight for 8 hr day	H, F, sex	Males/females based on psychophysical methodology
Lift, lower, push, pull,	Liles (1986)	Job Severity Index	Psychophysical capacity of working population, task characteristics	Based on psychophysical capacity of workers
Carry	Givoni and Goldman (1971)	Metabolic rate	L,S,G,W	Steady continuous carry on back, head, hands, and feet
Carry	Morrissey and Liou (1984)	Metabolic rate, heart rate	L,S,W	2 hands in front of body, continuous carriage
Carry	Randle *et al.* (1989)	Oxygen uptake, heart rate	L,D,F	2 hands in front of body, intermittent carriage
Lift, push/pull	Univ. of Michigan (1986)	Torque at: shoulder, elbow, hip, knee, ankle, low back	Force magnitude and direction, posture, anthropometric details	Sagittal plane static bio-mechanical model
Lift + carry, lower + carry	Taboun and Dutta (1984)	Oxygen uptake	F,L,W,D	Lift/carry, lower/carry
Lift, lower, carry or drag	Aberg *et al.* (1968)	Oxygen uptake	W, vertical and horizontal displacement	Requires centre of gravity measurement

Key
D = Distance of carry
F = Frequency of lift/carry
Fmax = Maximum lifting frequency
G = Gradient of surface
H = Height of lift
Ho = Height above floor of origin of lift/lower
Hd = Height above floor at destination of lift/lower
L = Load weight
S = Walking speed
W = Body weight
X = Horizontal movement of workpiece

have often been formulated without proper consideration of the many interacting components in the system. An examination of the system components in manual handling is therefore an important starting point for examining the functional nature of the guidelines.

A systems approach to manual handling

Before a systems approach for manual handling can be undertaken, the variables listed in Table 12.2 and their interactions should be considered. It is clear that no one model or set of guidelines can take all of these factors into account. Table 12.1 outlines the parameters which have been integrated into existing models. It is apparent that there is some consensus as to what should be included since load weight and frequency of handling are usually considered. However, it is not clear why these particular factors have been chosen when there is only limited scientific evidence to support their inclusion. It may be, therefore, that the most important determinant of what is and what is not included is the ease with which parameters can be controlled within a laboratory environment. In any event, it is rare to see such an environment in industrial workplaces.

A comparative study of some guidelines

Objective comparisons of all the guidelines are not available. However, one comparative study of data for bimanual lifting in the sagittal plane has been undertaken (Nicholson, 1986). This examined the guidelines developed using psychophysical criteria (Snook, 1978; Ciriello and Snook, 1983), force limits developed from intra-abdominal pressure (IAP) measurements (Materials Handling Research Unit, 1980) and those generated from a static sagittal plane lifting model (Chaffin and Andersson, 1984).

A number of limiting conditions were identified. One important distinction found between the guidelines was the concept of 'safe' and/or 'acceptable' manual handling forces. Nicholson (1989) has defined these in the following way:

> ACCEPTABLE limits to manual handling—derived from what a worker is willing to handle
> SAFE limits to manual handling—based on biomechanical and physiological criteria which it is considered unsafe to exceed.

Thus psychophysically determined guidelines propose *acceptable* weights of lift, lower, push, pull or carry—although some physiological validation has been obtained. Both infrequent and repetitive handling activities have been investigated. 'Force limits' and biomechanical models propose weights or forces which should be interpreted as *safe*—although little or no physiological validation has been performed. This is because the primary concern has been in limiting low back compressive force.

Nicholson (1989) found that the 'force limits' data proposed weights which were lower than those which are psychophysically acceptable, except for those tasks where lifting takes place close to and around the shoulder. The difference when lifting from below knee level was found to be in excess of 50%. Lifting the 'force limits' weights was found to generate compressive forces at L5/S1 which were some 55% below the NIOSH Action Limit (NIOSH, 1981), whereas lifting weights selected psychophysically was found to generate compressive forces about 10% in excess of the NIOSH Action Limit when lifting from the floor, but between 25% and 60% less when lifting from above knee level depending on the point of origin of the lift.

Table 12.2 Variables in manual handling systems

Component	Variable
Task characteristics	Location/position of object
	Frequency
	Duration
	Precision
	Type of handling activity
	Velocity
	Acceleration
	Time constraints
	Pay structure/incentive schemes
	Pacing–machine or self
	Work/rest cycles
	Shiftwork
	Job rotation
	Availability of assistance (teamwork)
Environment	Size and layout of workplace, e.g. access/egress, headroom, obstructions to free movement or reach
	Terrain/floor surface, e.g. steps, stairs, slopes, uneven ground, surface contaminants
	Temperature
	Humidity
	Lighting
	Visibility
	Noise
	Operating medium, e.g. air, water, space
	Motion, e.g. vibration, transport (trucks, ships)
	Clean, dirty, odorous
Object	Nature–animate/inanimate
	Weight/mass/resistance to movement/static or in motion or transitions
	Size and shape, e.g. physical dimensions, restrictions to visibility, uniform, rigidity, stability
	Centre of gravity–fixed or moveable
	Physical/chemical hazards
	Mechanical status, e.g. damage, fault, lack of repair
	Handling interface, e.g. handles/cutouts, surface contaminants, temperature
	Information/instruction, e.g. legibility, unknown language
Handler	Sex
	Age
	Strength/endurance
	Health status, e.g. sight, hearing
	Physical status, e.g. fitness
	Motivation
	Skill/knowledge
	Perception, e.g. physical, cognitive
	Anthropometric/anatomical characteristics
	State of adaptation, e.g. circadian rhythms
	Handedness
	Interaction with protective equipment

On comparing psychophysically acceptable bimanual pulling and pushing forces with those developed from the IAP methodology, it was found that there was a close agreement with the pushing data but a marked disagreement with the pulling data. The recommended 'force limits' data for pulling were in excess of other published limits except for those performed on high traction surfaces.

There are, therefore, areas of agreement and of difference between the guidelines. The provision of clear recommendations on suitable manual handling forces for work tasks or workstations can be particularly difficult. This is illustrated by reference to three case studies carried out by the Ergonomics Research Unit, University of Surrey, UK. Each case considers only one task within each job and does not attempt to consider the job in its entirety.

Case studies

Case study 1—lifting

A redevelopment of baggage handling facilities at Heathrow Airport, London, had led to concern among safety, management and union personnel that a potential back injury hazard had been created (Nicholson and Clark, 1986). The cause of concern lay in the design of the newly installed conveyor system. This new system was also to be the design for all future baggage handling lines. A vertical lift of 53 cm from the base of the incoming baggage container was necessary to place the baggage on the conveyor. With the old system the lift distance was only 4 cm. Management wanted to know:

- whether handling baggage from containers onto the new conveyor was likely to lead to back injury for the operatives; and
- whether the existing design was preferable in this respect.

On evaluation the 'force limits' data were found to be not applicable because they placed emphasis on the weight of the load and not on the lifting distance. In a similar way it was found that although distance of lift was a factor in determining the psychophysical acceptability of the load handled, the lift distance investigated in the studies (25, 51 or 76 cm) did not correlate with those in this application, and in any case the difference in recommended weight over the lift distances was only marginal. This precluded the use of the psychophysical data.

For these reasons, the Job Severity Index (JSI, after Liles, 1986) and the NIOSH equation (NIOSH, 1981) were used. A major difficulty with using these methods was the lifting techniques used by the baggage handlers. The JSI defines the worker lifting capability as the maximum acceptable weight of lift in the sagittal plane, and although it is stated that the JSI criterion is based on a large data base of mixed plane lifting activities, the preponderance of non-sagittal plane lifting in this case placed severe limitations on the interpretation of the analysis. Similarly, the NIOSH equation applies only to bimanual symmetrical lifting in the sagittal plane. Thus, whilst the calculations showed that the design of the existing conveyor was to be preferred, it was extremely difficult to advise management with any degree of certainty as to whether handling baggage on to the new conveyor was likely to lead to back injury for the operatives. In the event, management decided to revert to the existing system even though evidence from the study questioned the effectiveness of this design from a manual handling perspective.

Case study 2—pushing/pulling

The management of a large warehousing and distribution operation wished to know the maximum safe weight of supplies cages handled by the distribution drivers in order to minimise the risk of back injury. A study was undertaken in which the force applied at the hands and the load at the low back were measured while cages loaded to known weights were manoeuvred according to a set procedure. Details are published elsewhere (Nicholson *et al.*, 1986).

The data were compared with recommended maximum pushing and pulling forces in order to provide the guidance requested by management. A number of different reference sources were referred to in this context including psychophysical data and the 'force limits' data, but a lack of consensus was found. In addition, hand height, position of the feet relative to the load and the coefficient of friction between the shoe and the floor have a major bearing on the magnitude of acceptable pushing and pulling forces. Although training could be of assistance in this regard, the drivers and management have no control over the environment in the customer's premises through which the loaded cages have to be transported. Thus, although it was possible to advise management on an appropriate loaded cage weight (despite there being some lack of consensus in the reference sources), it was also necessary to stress that the recommended weight was only applicable to optimum loading and unloading conditions. This placed limitations on the practicability of the recommendation in working environments.

Case study 3—carrying

The authors have encountered a number of situations in which the carriage of loads, rather than lifting or pushing/pulling, was the predominant activity. In the majority of these cases it has not been possible to use prediction models or load carriage guidelines, due to their lack of applicability to the specific task under consideration. It was therefore necessary to make experimental assessments of the acceptability of the tasks.

For example, in one study we were asked to evaluate the physical strains incurred in carrying a large rectangular container (54 kg) for up to 2 km over rough terrain. The container had a handle on either side, and was designed to be carried by two people (i.e. a one-handed carry). Of the models investigated, only the metabolic rate prediction model of Givoni and Goldman (1971) was applicable for walking over rough terrain. However, this was primarily intended for load carriage on the back and was not appropriate for carrying a heavy load in one hand only.

Similar problems were encountered in two other studies involving delivery drivers. One involved the collection and delivery of television sets (up to 38 kg) and the other the handling of medical waste containers (up to 25 kg). In both situations the most frequent and strenuous part of the load carriage task involved the negotiation of staircases. In many cases these were very steep, twisting or obstructed. None of the available guidelines or models was applicable to carrying up or down staircases, most being based on the responses to walking on a level surface or on a treadmill.

Conclusion

The case studies have shown the difficulties of applying the models to industrial situations. In each case, none of the guidelines was wholly appropriate. This raises the question of whether it is better to discount their use or to apply them but only use the results as a vague approximation of the true load. The dilemma may, in some instances, be resolved through careful consideration of the assumptions made in developing each of the models and through a detailed assessment of the factors (see Table 12.2) which may be present. It is apparent that any one of these factors (e.g. handling very hot objects) or, more probably, that a combination of several factors (e.g. handling objects with poor handles, with a low coefficient of friction and in restricted headroom) may preclude the use of the guidelines. In addition, some major discrepancies have been found between maximum recommended force applications. It is not clear whether studies involving more task parameters would improve this situation.

In practice, and in the authors' experience, it seems that direct application of any of the guidelines currently available is rarely satisfactory. A better, but more expensive approach, is to consider each task and, where possible, the interaction of task variables in a systematic way. In addition, it may be necessary to support these with an experimental evaluation of the tasks 'in situ', and where necessary in a laboratory simulation. This has generally led to more acceptable recommendations although their true long-term effectiveness is difficult to assess.

The existing models are perhaps only of real use to experts in the area who have an understanding of their limitations and underlying assumptions. Those without such an understanding should use them with caution, both for existing tasks and in relation to the design of new working systems. Advice should be sought if doubts over applicability exist. Such an approach is likely to be advocated in forthcoming guidance and directives from both the UK Health and Safety Commission and the EEC.

References

Aberg, U., Elgstrand, K., Magnus, P. and Lindholm, A. (1968). Analysis of components and prediction of energy expenditure in manual tasks. *International Journal of Production Research*, **6**, 189–196.

Asfour, S.S. (1980). *Energy cost prediction models for manual lifting and lowering tasks*. PhD thesis, Texas Technical University, Texas, USA.

Chaffin, D.B. and Andersson, G. (1984). *Occupational Biomechanics*. (New York: John Wiley).

Ciriello, V.M. and Snook, S.H. (1983). A study of size, distance, height and frequency of effects on manual handling tasks. *Human Factors* **25(5)**, 473–483.

Garg, A., Chaffin, D.B. and Herrin, G. (1978). Predictions of metabolic rates for manual materials handling jobs. *American Industrial Hygiene Association Journal*, **39**, 661–674.

Givoni, B. and Goldman, R. (1971). Predicting metabolic energy cost. *Journal of Applied Physiology*, **30**, 429–433.

Karwowski, W. and Ayoub, M.M. (1984). Effect of frequency on maximum acceptable weight of lift. In *Trends in Ergonomics/Human Factors* **I**, edited by W. Karwowski, (Amsterdam: Elsevier), pp.169–172.

Liles, D.H. (1986). The application of the job severity index to job design for the control of manual materials handling injury. *Ergonomics*, **29**, 65–76.

Materials Handling Research Unit (1980). *Force limits in manual work*, (London: Butterworth).

Mital, A., Shell, R.L., Mital, C., Savghavi, N. and Ramanann, S. (1984). *Acceptable weight of a lift for extended work shifts*. Tech. Rep. No. 1-R01-0H-01429 01/2. (Cincinnati: NIOSH).

Morrissey, S. and Liou, Y. (1984). Metabolic cost of load carriage with different container sizes. *Ergonomics*, **27**, 847–853.

Nicholson, A.S. (1986). Manual handling limits—a comparative study. In *Trends in Ergonomics/Human Factors* **III**, edited by W. Karwowski, (Amsterdam: Elsevier). 793–800.

Nicholson, A.S. (1989). A comparative study of methods for establishing load handling capabilities. *Ergonomics*, **32**, 1125–1144.

Nicholson, A.S. and Clark, A.G. (1986). The use of handling guidelines in assessing baggage handling facilities. In *Trends in Ergonomics/Human Factors* **III**, edited by W. Karwowski, (Amsterdam: Elsevier), pp.855–862.

Nicholson, A.S., Ridd, J.E. and Fernandes, A.F. (1986). Handling problems associated with the distribution of supplies. In *Contemporary Ergonomics 1986*, edited by D.J. Oborne, (London: Taylor & Francis), pp.232–236.

NIOSH–National Institute for Occupational Safety and Health. (1981). *Work handling*. DHHS/NIOSH Publication no. 81–122, (Cincinnati, OH: NIOSH).

Randle, I., Legg, S. and Stubbs, D. (1989). Task based prediction models for intermittent load carriage. In *Contemporary Ergonomics 1989*, edited by E. Megaw, (London: Taylor & Francis), pp.380–385.

Snook, S.H. (1978). The design of manual handling tasks. *Ergonomics*, **21**, 963–985.

Taboun, S. and Dutta, S. (1984). Prediction models for combined tasks in manual handling. *Proceedings of International Conference on Occupational Ergonomics, Toronto, Canada*, pp.551–555.

University of Michigan (1986). *2D static strength prediction program*. The Center for Ergonomics, University of Michigan, Ann Arbor, Michigan, USA.

Chapter 13
Computer-aided ergonomic design: current status

Arto Kuusisto
Tampere University of Technology, Tampere, Finland

Introduction

Traditional design of workplaces seldom takes ergonomics parameters into consideration. It seems that either the existing ergonomics data are not available or workplace designers do not consider it to be very important. This situation is not changing despite the fact that several national and international laws require designers to use existing ergonomics standards and directives. New workplaces are often designed using computer-aided design (CAD) systems. These systems can relatively easily handle human performance information, and could therefore provide ergonomics data for designers at an early design stage.

Structure of the present ergonomic design programs

Since the end of the 1960s several dozen computer programs have been created to assist ergonomic workplace design. These programs have been used mainly in the aircraft and vehicle industries. The first ergonomics computer programs were based on geometric man-models, and permitted user specification of the man-model's anthropometry and posture. In the beginning of the 1980s, some biomechanical calculations were included in order to evaluate the worker's physical load (Dooley, 1982).

Current ergonomics programs (Figure 13.1) usually comprise two parts: (1) work space evaluation and (2) assessment against biomechanical criteria. Work space evaluation deals with the interaction between man and his environment. The biomechanical assessment involves evaluation of work postures (often using methods like OWAS (Karhu *et al.* 1977)) and determination of the physical work capacity. Static strength can be calculated by using methods such as those devised by Chaffin and Andersson (1984).

Work space evaluation	Biomechanics
Dimensions of the workplace	Posture
Visual field	Static load
Reach and comfort area	Dynamic load

Figure 13.1. The basic structure of the current ergonomics programs

Some programs have additional features not listed here, such as the CAR program (Dooley, 1982) which was created only for reach analysis, and the 2-D model from the University of Michigan is purely biomechanical (University of Michigan, 1986). Text files are rarely used in these programs, although they could be useful in displaying information which is difficult to present graphically. Text files could also be used to provide rapid checklists during the design process.

Modern ergonomics programs either function as self-contained software or are integrated into a CAD system, such as SAMMIE (Kingsley *et al.*, 1981). The 2-D static strength program from the University of Michigan is an example of a self-contained program. Ergonomics data from such programs can be transmitted to a design system using data exchange files, but this method is time-consuming and not effective in routine design. Most man-model programs are three-dimensional and can be used only on minicomputers. OSCAR is one of the few 3-D models which can be run on an IBM-compatible microcomputer (Lippman, 1986), and ErgoSHAPE is a 2-D man-model program which runs on an IBM microcomputer (Launis *et al.*, 1988).

Problems with the current computer-aided ergonomic design

Some aspects of computer-aided design were not apparent when the current ergonomics programs were created. In the 1970s and early 1980s it was commonly expected that CAD systems would rapidly replace manual drawing and design methods. It was also thought that various subprograms (e.g. for ergonomics) would come into use within CAD. It is now obvious, however, that this was overly optimistic. Traditional design methods are still being used, and CAD is used mostly for workshop applications. The use of computer-aided human workplace design is often limited to pictures of layouts (Launis *et al.*, 1989).

Lack of user-friendliness is a problem common to many 3-D man-model programs, and prevents their use in industrial design applications. Difficulty is caused, for example, by determination of the various joint angles when the working posture is changed. Models involving more than twenty joints are difficult to handle, especially if the program has no automatic functions to specify movements of the man-model (e.g. reach). Because of these factors, the 3-D man-models are simplified, and work space dimensioning can only be accomplished at an approximate level. In order to improve their usability, future ergonomics programs should contain the following features:

- quick to use at every design stage,
- equipped with simple and effective interaction,
- usable without any previous experience of ergonomics,
- able to give reliable design parameters, and
- equipped only with features really needed by the designer.

Expert systems in workplace design

Expert systems, or knowledge-based systems as they are often called, can assist a design engineer in achieving an ergonomic workplace design. These systems have only recently been developed and can solve problems requiring human judgement or expertise due to the uncertainty or complexity associated with the design tasks. Current expert systems are confined to well-circumscribed tasks. They always attend to detail, and they always consider all the possible alternatives systematically (DeGreve and Ayoub, 1987). Workplace design is a problem domain which is relatively narrow, yet sufficiently unstructured and uncertain to merit the development of a dedicated expert system. A workplace design expert system has the potential to utilize guidelines from the literature, to borrow experience from a human expert, and to apply this knowledge at a practical level (DeGreve and Ayoub, 1987).

Occupational health care computer systems

In Finland, employers are obliged to provide occupational health care (OHC) for employees. The main task of OHC personnel is nursing, but in Finland they also perform systematic workplace investigations. Major companies have a health care system of their own, and in these the workplace investigations are performed more frequently and thoroughly. However, about 50% of Finnish workers use an OHC system from outside the company, and in these cases the workplace investigations have been less effective. OHC personnel collect a great amount of data dealing with the workers' health and work environment, but at present these data are used too seldom by workplace designers. The diseases and injuries are recognized, but this recognition leads to few changes in workplaces.

The OHC personnel have developed some systematic methods for performing workplace investigations, one of which is Job Load and Hazard Analysis (Mattila, 1987). Recently the OHC personnel have also obtained computer programs to improve their capabilities. These programs contain modules for:

- reservations
- accident statistics
- data from visits to the medical doctor
- education data
- work environment data
- financial data
- reports
- office automation
- outside links

Most OHC computer systems have a module for collecting the work environment data, but to date this module has been used very little, mainly because of the lack of methods for evaluating the data which is collected. Expansion of this module would give a database not only for OHC personnel, but also for safety personnel, workers and workplace designers and redesigners (Figure 13.2). It would be possible to use the data collected from the workplaces together with design instructions from the literature and handbooks to develop an expert system which can give direct guidelines for improved ergonomic design of workplaces.

Conclusions

There are now some modern ergonomics programs which can be used for routine design tasks. Some features in these programs have prevented their widespread use in workplace design. In particular, poor compatibility with the main drawing program, lack of user-friendliness and high price are common problems.

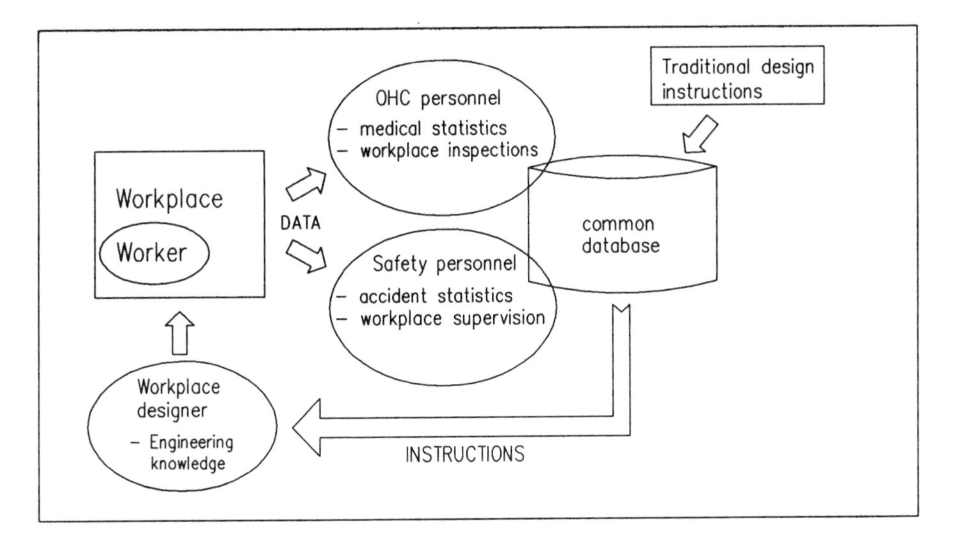

Figure 13.2. A common database at the disposal of groups having an effect on workplace design.

Countries where occupational health care is provided have the prerequisites for systematic workplace investigations, and an expert system using the data collected in workplace investigations would give the designer task-oriented design instructions. The main reasons why expert systems for ergonomic workplace design should be developed and used are:

- an expert system can produce direct design instructions for the designer,
- the designer gets current data (feedback) directly from the existing workplace, and
- cooperation between OHC personnel, safety personnel, workers and designers improves.

The role of safety personnel is particularly important when workplaces are *redesigned*. Because the safety personnel usually include people with a technical education, they are likely to be able to identify deficiencies and hazards of the environment more easily than the OHC personnel. Accident statistics are usually kept by safety personnel, and these data should be available in a common database in order to recognize the most hazardous workplaces and tasks.

Finally, in the initial phases of the process of designing a new workplace, it is highly important that all groups—designers, safety personnel, occupational health care personnel and workers—should work together from the very beginning.

References

Chaffin, D. and Andersson, G.B.J. (1984). *Occupational Biomechanics*, (New York: Wiley & Sons).

DeGreve, T.B. and Ayoub, M.M. (1987). A workplace design expert system. *International Journal of Industrial Ergonomics*, **2**, 37–48.

Dooley, M. (1982). Anthropometric Modeling Programs—A Survey. *IEEE Computer Graphics & Applications*, **9**, 17–25.

Karhu, O., Kansi, P. and Kuorinka, I. (1977). Correcting Working Postures in Industry:

A practical method for analysis. *Applied Ergonomics*, **8**, 199–201.
Kingsley, E.C., Schofield, N.A. and Case, K. (1981). Sammie—Computer-aid for Man-Machine Modeling, *Computer Graphics, (Proc. Siggraph'81)*, **15**, 163–169.
Launis, M., Lehtelä, J. and Kuusisto, A. (1988). Kaksiulotteinen antropometrinen ihmismallijärjestelmä työpaikkojen suunnitteluun (CAD). (2-dimensional anthropometric man-model system for the design of workplaces). *Työ ja ihminen*, **1**, 68–83. (With an English summary.)
Launis, M., Lehtelä, J., Haajanen, M. and Ahonen, M. (1989). *Ergonomisen tietokonesuunnittelun käyttömahdollisuudet ja edellytykset työpaikkojen fyysisessä suunnittelussa*. (The usability and qualifications of computer aided ergonomic design in the design of work environments). Työsuojelurahaston tutkimusraportti No 87034, (Helsinki: Institute of Occupational Health).
Lippman, R. (1986). Arbeitsplatzgestaltung mit Hilfe von CAD. *REFA-Nachrichten*, **3**, 13–16.
Mattila, M. (1987). Job load and hazard analysis: a method to identify job safety problems and to produce preventive measures. In *New Methods in Applied Ergonomics*, edited by J.R. Wilson, E.N. Corlett and I. Manenica (London: Taylor & Francis), pp. 199–204.
University of Michigan (1986). *User's manual for the two dimensional static strength prediction program*. Center for Ergonomics, The University of Michigan, USA.

Chapter 14
Job evaluation procedures to identify ergonomic stress

W. Monroe Keyserling and Thomas J. Armstrong

Department of Industrial and Operations Engineering,
Center for Ergonomics, The University of Michigan,
Ann Arbor, Michigan 48105, USA

Introduction

Over-exertion injuries resulting from work activities (e.g., low back pain, cervicobrachial disorders, and upper extremity disorders) are a major public health problem in virtually all industrialized nations (Armstrong, 1986; USDHHS, 1987; Jonsson *et al.*, 1988). Many of these problems are associated with excessive physical demands due to poor design of facilities, equipment, tools and work methods. A principal objective of many industrial ergonomics programs is to reduce or eliminate the exposure of workers to excessive stress through the process of job re-design. Job analysis is an integral part of the re-design process because it provides a methodology to answer the following questions:

- Recognition: Does performing the job expose workers to conditions (i.e., risk factors) which are likely to cause injury?
- Diagnosis: What attributes of the job cause or contribute to these exposures?

These questions must be answered as a first step in the re-design process.

The methodology described below presents a structured approach to ergonomic job analysis which is similar to the Systems Hazard Analysis techniques developed by safety engineers (Firenze, 1978). This requires the development of a detailed description of work place features and job activities. Each step of the job is evaluated to determine the presence of generic risk factors which may result in over-exertion injuries.

Methodology

Step I—job documentation

The purpose of job documentation is to develop a complete description of how the job is currently performed. This description serves two major functions: (1) to assist the analyst in identifying and diagnosing risk factors, and (2) to establish baseline measurements which can be used to evaluate the effectiveness of changes resulting from job re-design. Documentation involves the following items.

Work objectives

This is a very short (one or two sentence) description of the major purpose of the job (e.g., what the worker is expected to produce or accomplish).

Work schedule

This information is used to determine a worker's total daily exposure to the risk factors associated with the job. The purpose is to establish the amount of time (typically measured in minutes or hours) for which the job is performed.

Worker characteristics

A worker's attributes (e.g., height, weight, dominant hand, work experience, etc.) may affect the way he/she performs the job.

Production information

Production information, such as the speed of an assembly line or time standards, is needed to determine the number of work cycles performed per shift. These data, used to estimate the repetitiveness of a job, can often be obtained from industrial engineering records. In facilities which use incentive pay programs, production information can be obtained from payroll records.

Job rotation schedule

In some plants, workers rotate among different jobs on a regular basis. Although a job rotation plan does not change the ergonomic risk factors associated with a specific job, it does affect the cumulative exposure experienced by the worker. If job rotation is used, it is important to record the amount of time spent on each job.

Work station layout

A work station sketch should be drawn which shows the location of all work benches, equipment, conveyors, supply bins, etc. This sketch should include horizontal dimensions for walking distances and required reaches. Vertical dimensions should also be recorded as many awkward postures result from work locations that are either too high or too low (Armstrong, 1986; Keyserling *et al.*, 1987).

Equipment and tools

These descriptions do not have to be detailed. For hand-held tools, however, basic information such as tool weight and/or torque settings can be used to estimate forces required to operate the tool.

Parts and materials

All parts and materials which are manually handled should be listed. Descriptive information, such as the weight and external dimensions of the part, should also be recorded.

Work methods

Each ergonomic problem associated with a job should be associated with a specific operational step or work element. Most jobs involve a combination of regular elements (which are typically performed on each production unit and follow a fixed sequence) and irregular elements (typically performed a few times per work shift). It is sometimes useful to record the duration and frequency of regular work elements.

Step II—identification of generic risk factors

Each work element should be examined to determine if it results in exposure to one of the generic risk factors described below. If excessive levels of exposure are found, the contributing causes should be described. Detailed procedures for evaluating the nature and magnitude of ergonomic stresses are beyond the scope of this chapter. It is assumed that at least one member of the job analysis team will have sufficient ergonomics skills to perform the necessary analyses.

Forceful exertions

Forceful exertions such as 'whole-body' manual materials handling tasks (e.g., lifting, pushing, etc.) can result in back pain and other musculoskeletal injuries (NIOSH, 1981). In general, heavy loads require forceful exertions and are associated with increased risk of injury. Due to the effect of long moment arms, relatively light loads may also require forceful exertions (and increased risk of injury) if the load is horizontally displaced from the spine.

Forceful exertions of the hands (e.g., using knives, wrenches, and other hand tools, using the fingers and hands to shape materials and parts, etc.) can cause upper extremity cumulative trauma disorders such as inflammation of joints, muscles, and tendons as well as diseases of the peripheral nerves (Welch, 1972; Armstrong, 1986). Heavy tools, poorly-balanced and/or poorly-maintained tools (e.g., dull knives and scissors) increase the force requirements of manual work. Gloves may also increase the muscular effort required to perform gripping tasks (Hertzberg, 1955).

Awkward postures

Awkward trunk postures such as flexion, lateral bending, and twisting increase the likelihood of back injuries (Keyserling *et al.*, 1987), particularly during lifting (NIOSH, 1981). Excessive shoulder elevation may increase the likelihood of cervicobrachial disorders (Jonsson *et al.*, 1988). Most of these postures are associated with poor work station layout which requires excessive reach distances.

Seated work may also contribute to the development of back pain. When a person adopts a seated posture (trunk-thigh angle of about 90 degrees), the lumbar spine flattens producing increased stresses on the spinal discs, ligaments and muscles (Andersson *et al.*, 1979; Yu *et al.*, 1988). Clinical and epidemiological studies have found increased rates of low back pain among workers who sit for prolonged periods (Magora, 1972; Kelsey and Hochberg, 1988). The design of the work seat (e.g. the angle of the backrest relative to the seat, the use of a lumbar support, and the angle of the seatpan) and the physical requirements of the job are related to the level of biomechanical strain (Bendix, 1984; Yu *et al.*, 1988).

Upper extremity cumulative trauma disorders are associated with pinching and deviated wrist postures such as excessive flexion, extension, radial deviation, and/or ulnar deviation (Armstrong 1986). Awkward wrist postures may be caused by workstation layout and/or the features of hand tools such as the shape and orientation of handles. Elbow disorders are associated with extreme pronation and/or supination of the forearm, particularly if these motions are highly repetitive.

Mechanical stresses

Mechanical stresses are caused by physical contact between body tissue and an object or tool. These stresses are associated with work activities where a body part is in contact with a hard or sharp object and/or when a body part is used as a striking tool. Forceful gripping of tools with sharp handles can produce localized pressure on the underlying tendons in the fingers, resulting in tenosynovitis and related disorders. Tools with ringed handles which rub against the side of the fingers (e.g., scissors) can traumatize the digital nerves. Tools and other objects which are supported by the base of the palm can produce pressure on the median nerve and contribute to the development of carpal tunnel syndrome (Tichauer, 1966; Armstrong, 1986).

The seatpan of a chair may exert sufficient localized force on the posterior aspect of the thigh to produce extreme discomfort and in some cases impair circulation to the lower extremities. Factors such as the size, contour, and slant of the seatpan affect the location and magnitude of mechanical stresses (Yu *et al.*, 1988).

Vibration

Whole body vibration, frequently associated with the driving of motor vehicles, is associated with elevated rates of low back pain (Frymoyer *et al.*, 1980; Kelsey and Hochberg, 1988). Since driving tasks are usually performed in a sitting posture, most drivers are exposed to two back pain risk factors when performing their jobs. Driving over rough surfaces for prolonged periods and vehicle seats with poor suspension systems increase exposure to whole-body vibration.

Localized vibration of the upper extremities may occur during work activities that involve the use of powered tools (e.g., chainsaws, pneumatic wrenches, jackhammers, etc.). Vibration has been shown to increase grip force exertions in a laboratory simulation of hand-tool use (Radwin *et al.*, 1987). Vibration may also be a contributing factor to the development of carpal tunnel syndrome and vascular disorders (Taylor and Pelmear, 1975; Armstrong, 1986).

Temperature extremes

The risk factors discussed thus far can usually be associated with specific work elements. Exposure to unusually hot or cold temperatures may result from the thermal characteristics of the general work environment (e.g., a room that is hot or cold) or from specific work elements (e.g, handling hot items which have just been removed from an oven, handling cold parts and/or tools).

Working in a hot environment can lead to heat stress and fatigue. Working in a cold environment, handling cold parts and/or exposing the fingers to cold exhaust from a pneumatic tool can reduce manual dexterity (Armstrong, 1986). While gloves may be used to protect the hands and fingers from cold temperatures, they may increase the force required to perform a task as discussed above.

Repetitive or prolonged activities

The ergonomic strain experienced by a worker is related to his/her cumulative exposure to the risk factors discussed above. Biomechanical and physiological strain increases with repetitive or prolonged exposure to forceful exertions, awkward postures, local mechanical stresses, vibration and/or temperature extremes. In some cases, repetitiveness in the absence of other factors may increase the risk of injury (Silverstein, 1985).

Psychophysical studies of manual materials handling have demonstrated that the maximum acceptable force during lifting, pushing and pulling is inversely related to the repetitiveness of the task (Snook, 1978). Lifting frequency has been cited by the *NIOSH Work Practices Guide for Manual Lifting* (1981) as a risk factor contributing to fatigue and low back pain during lifting activities. Prolonged work in non-neutral trunk postures (e.g., forward flexion, lateral bending, or twisting) has also been associated with back pain (Keyserling *et al.*, 1987).

Repetitiveness is commonly cited as a risk factor associated with the development of upper extremity cumulative trauma disorders in hand-intensive jobs (Hymovich and Lindholm, 1966; Tichauer, 1966; Silverstein, 1985; Armstrong, 1986). It is difficult to develop a precise definition of repetitiveness. However, repetitiveness can be estimated in a number of different ways. For example, on an assembly line repetitiveness is related to the time allowed to complete one unit of work. In this case, repetitiveness is a function of the production rate. Jobs which have a basic cycle time of 30 seconds or less (i.e. a production rate of two or more parts per minute) have been found to have elevated rates of upper extremity cumulative trauma disorders (Silverstein, 1985). In other cases, repetitiveness can be independent of the cycle time. Any job that involves repeated motion patterns and/or prolonged postures within a work cycle may be considered repetitive. In these instances, repetitiveness can be described as the number of motions per unit of time or the percentage of working time spent in certain postures.

The job analysis team

In order to document and analyze the job properly, it is necessary to assemble a job analysis team. The size and membership of this team will vary depending on the organization of the ergonomics program and the nature/complexity of the job. It is recommended, however, that the team always include the following members:

- the *worker* who performs the job. This person can provide detailed descriptions and explanations of the work method and provide insights to the nature and causes of injuries, disorders, fatigue, and discomfort.
- the *supervisor*. Like the worker, the supervisor is an excellent resource for documenting the work method and identifying the nature and causes of ergonomic stress. The supervisor may also provide historical information as to how the job has affected people who no longer perform the job.
- the *ergonomist*. This person has been trained in the recognition and assessment of the generic risk factors described above. His/her role is to work with the team to identify job attributes which cause ergonomic stress, and to evaluate the magnitude of the hazard.

It is frequently beneficial to include other persons in the job analysis team. The *plant physician, nurse* or *safety manager* is familiar with the nature and severity of injuries and disorders associated with the job. *Engineers* can provide technical information concerning the magnitude of risk factors (e.g., the hand force required to tighten a bolt is determined by the torque specifications for the joint). Engineers are also involved in evaluating the feasibility of process changes to reduce risk factors. Any other person who can provide insight to the causes and/or control of ergonomic stress should be included in the team.

Summary

Ergonomic job analysis is an open-ended process that involves detailed inspection, description and evaluation of the workplace equipment, tools and work methods. The method described above provides a structured approach for performing job analyses. Table 14.1 summarizes some of the common ergonomic risk factors found in industry. This checklist, used in combination with a detailed description of work activities, can assist the job analysis team in the identification of generic risk factors.

Table 14.1 Risk factor checklist for ergonomic job analyses

Risk factor	Associated work activity	Comments
Forceful exertions		
Lifting/materials handling		
Upper extremity		
Heavy/poorly balanced tool		
Poorly maintained tool		
Grinding/shaping		
Awkward postures		
Kneeling		
Prolonged standing		
Prolonged sitting		
Trunk bending/twisting		
Neck bending/twisting		
Elevated shoulder		
Rotated forearm		
Deviated wrist		
Mechanical stresses		
Sharp edges/workbench		
Sharp edges/handtool		
Ringed handles		
Hand hammering		
Seat Shape/padding		
Vibration		
Whole body		
Localized		
Temperature extremes		
Hot work area		
Cold work area		
Hand held tool/object		
Impermeable clothing		
Repetitive/prolonged activities		
Short cycle time		
Repeated/prolonged stress		
Trunk/lifting		
Shoulder		
Upper extremity		
Whole body (energy expenditure)		

References

Andersson, G.B.J., Ortengren, R., Nachemson, A. and Elfstrom, S. (1979). The influence of backrest inclination and lumbar support on lumbar lordosis. *Spine*, **4**, 52–58.

Armstrong, T. (1986). Ergonomics and cumulative trauma disorders. *Hand Clinics*, **2**, 553–565.

Bendix, T. (1984). Seated trunk posture at various seat inclinations, seat heights, and table heights. *Human Factors*, **26**, 695–703.

Firenze, R. (1978). *The process of hazard control*, (Dubuque, Iowa: Kendall/Hunt Publishing Co.).

Frymoyer, J., Pope, M., Constanza, M., Rosen, J. and Goggin, J. (1980). Epidemiologic studies of low back pain. *Spine*, **5**, 419–423.

Hertzberg, T. (1955). Some contributions of applied physical anthropometry to human engineering. *Annals of the New York Academy of Science*, **63**, 616–629.

Hymovich, L. and Lindholm, M. (1966). Hand, wrist, and forearm injuries, the result of repetitive motions. *Journal of Occupational Medicine*, **8**, 573–577.

Jonsson, B., Persson, J. and Kilbom, A. (1988). Disorders of the cervicobrachial region among female workers in the electronics industry. *International Journal of Industrial Ergonomics*, **3**, 1–12.

Kelsey, J. and Hochberg, M. (1988). Epidemiology of chronic musculoskeletal disorders. *Annual Review of Public Health*, **9**, 379–401.

Keyserling, W., Fine, L. and Punnett, L. (1987). Postural stress of the trunk and shoulders: identification and control of occupational risk factors. In *Ergonomic interventions to prevent musculoskeletal injuries in industry*. (Chelsea, MI: Lewis Publishers), pp. 11–26.

Magora, A. (1972), Investigation of the relation between low back pain and occupation; 3: physical requirements: sitting, standing, and weight lifting. *Industrial Medicine and Surgery*, **41**, 5.

National Institute for Occupational Safety and Health (NIOSH) (1981). *Work practices guide for manual lifting*, Technical Report No. **81–122** (Cincinnati, OH: NIOSH).

Radwin, R., Armstrong, T. and Chaffin, D. (1987). Power handtool vibration on grip exertions. *Ergonomics*, **30**, 833–855.

Silverstein, B. (1985). *The prevalence of upper extremity cumulative trauma disorders in industry*, Ph.D. Dissertation. (Ann Arbor, MI: University of Michigan).

Snook, S. (1978). The design of manual handling tasks. *Ergonomics*, **21**, 963–985.

Taylor, W. and Pelmear, P. (1975). *Vibration white finger in industry*. (London: Academic Press).

Tichauer, E. (1966). Some aspects of stress on the forearm and hand in industry. *Journal of Occupational Medicine*, **8**, 63–71.

United States Department of Health and Human Services (USDHHS) (1987). Conference on injury in America, *Public Health Reports*, **102**, 574–676.

Welch, R. (1972). The causes of tenosynovitis in industry. *Industrial Medicine*, **41**, 16–19.

Yu, C., Keyserling, W. and Chaffin, D. (1988). Development of a work seat for industrial sewing operations: results of a laboratory study. *Ergonomics*, **31**, 1765–1786.

Chapter 15
Ergonomic design of workplaces with a two-dimensional micro-CAD system

M. Launis and J. Lehtelä
Institute of Occupational Health, Helsinki, Finland

Background to the project

There is a basic contradiction in the ergonomic development of workplaces. In practice, ergonomists are usually restricted to modifying the existing workplaces, while design of new workplaces' is usually done without any ergonomic knowledge at all. The problem is multi-factorial, being connected to design goals, design organisation, design education and design practice, as well as to the quality of ergonomic knowledge and ergonomic design methods.

The ergonomic recommendations aimed for use in technical design are mostly detailed technical specifications which are tied to the existing technology or products, and are thus soon out-dated. The result is a heterogeneous recommendation practice, which creates confusion among the planners in the hasty design process. On the other hand, general knowledge about human functions and ergonomics is regarded as either trivial or too theoretical.

Ergonomic knowledge and methods can also be considered a means for activating expert cooperation and participation in planning. Cooperation between planners and experts in ergonomics, both in health care and safety organisations is needed, but in practice it has been occasional and ineffective. One of the reasons mentioned for this problem is the difficulty of reading technical drawings.

To create better ergonomic design tools the following criteria have been formulated (Launis *et al.*, 1988):

- they should be easy and fast to use, at exactly the right phase of the design
- they should be universally applicable to different problems and new situations
- they should be capable of tackling many design goals and tasks simultaneously, e.g. analysis, evaluation, synthesis and documentation
- they should be illustrative in order to facilitate cooperation between different experts and occupational groups
- they should have a clear and rational basis in order to be applicable to difficult or critical situations
- special training should not be required of the user, but the tools should also be suitable for expert consultations.

Two-dimensional anthropometric manikins and drawing templates have been introduced as a universal ergonomic design tool but this equipment has never been taken into wide design use, probably because of practical problems caused by different body sizes and various drawing scales. On the other hand, the sophisticated man-model programs available so far have been too expensive for everyday use. Computer-aided design (CAD) programs, especially the microcomputer based ones, are coming into wide use and offer new possibilities for solving the dilemma mentioned above.

ErgoSHAPE system

The ergoSHAPE system has been developed at the Institute of Occupational Health in Finland to provide ergonomic knowledge for the design of workplaces. The system is constructed inside a micro-CAD program called AutoCAD*. The ergoSHAPE system is two-dimensional, which means that the plans are drawn in perpendicular projections, as is usual in technical drawings. Although this is an incomplete representation of the three-dimensional real-life working situation, it has many practical advantages. Two-dimensional design is easy to perceive, easy to learn and use, and effective in normal routine design work. It also runs on a microcomputer and is therefore widely usable. In addition, two-dimensional man-models can be made anthropometrically more accurate and more realistic in their projections than the rough, geometrically simplified three-dimensional man-models available so far.

The ergoSHAPE system consists of three parts:

- anthropometric man-models, with the help of which working space can be fitted to human dimensions (Figures 15.1 and 15.3)
- recommendation charts, which give direct design guidelines for particular work situations and workplaces (Figure 15.2)
- biomechanical calculations, which enable the evaluation of postural stress in manual material handling or in static postures (Figure 15.4).

One hundred and eighty different versions of human models have been constructed. The reason for the great number of models is to ensure that the most suitable model can be found for any design purpose. The models combine the following characteristics (Figure 15.1):

- projection (viewing direction): left, right, top and front
- complexity (movability): fixed one-part models in typical working postures, models with six separate parts (trunk, head, two arm and leg segments), or nine parts (lower and upper trunk, head, three arm and leg segments)
- basic posture: sitting and standing, including some minor anthropometric differences
- sex: male and female
- anthropometric size: small, average and large (5th, 50th and 95th percentiles)

The models can be supplied with curves indicating the viewing angles and distances and the reach zones. The models and other data can be manipulated by built-in macro-commands which are displayed on the screen or on the digitising tablet. Thus the technical use of the ergoSHAPE system does not require any special training.

---

*AutoCAD is a registered trademark of Autodesk, Inc.

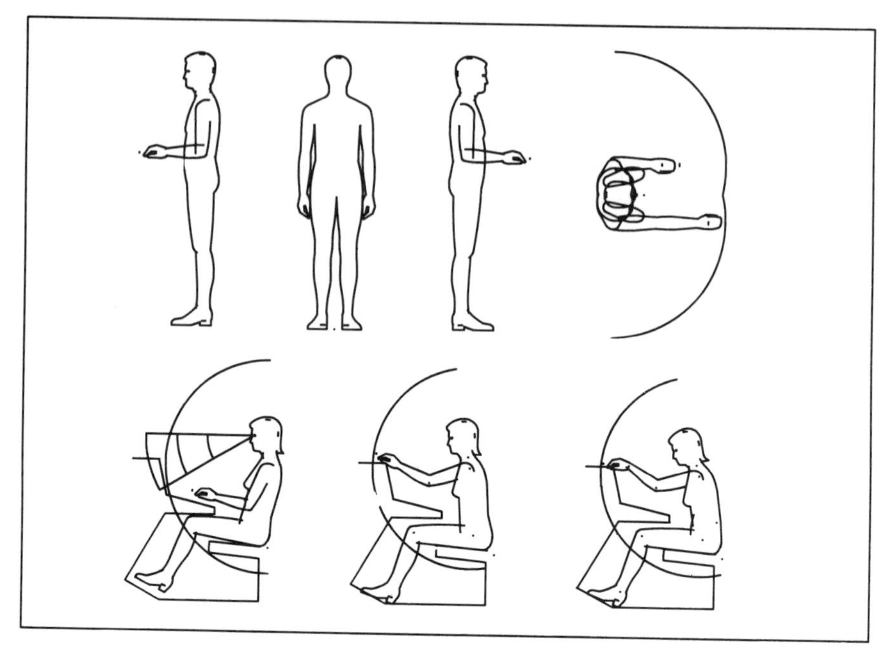

Figure 15.1. Examples of man-models of the ergoSHAPE system.
Topline: left, front, right and top projections.
Bottom line: fixed one-part model and movable six- and nine-part models.

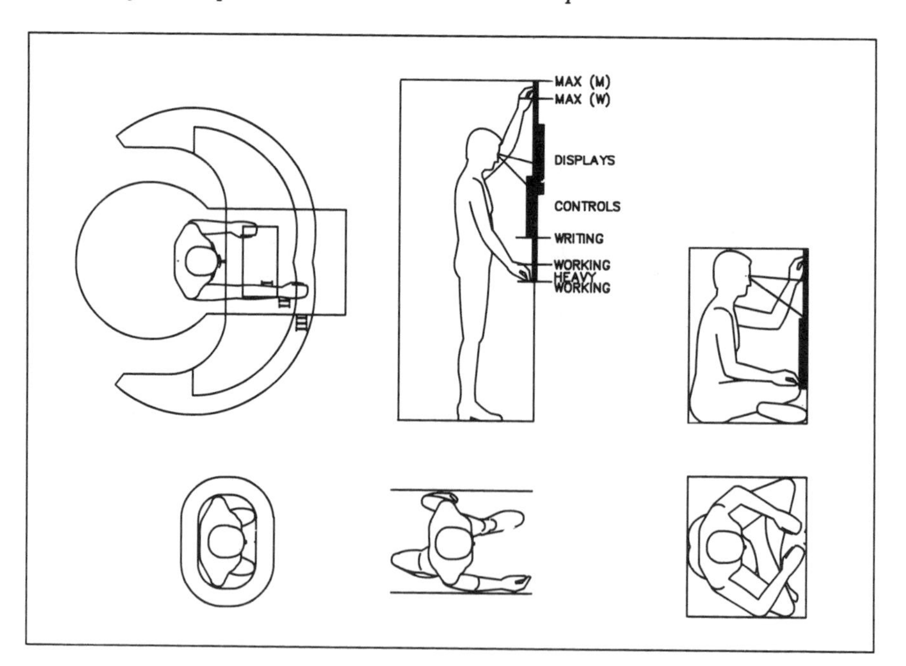

Figure 15.2. Examples of recommendation charts of the ergoSHAPE system.

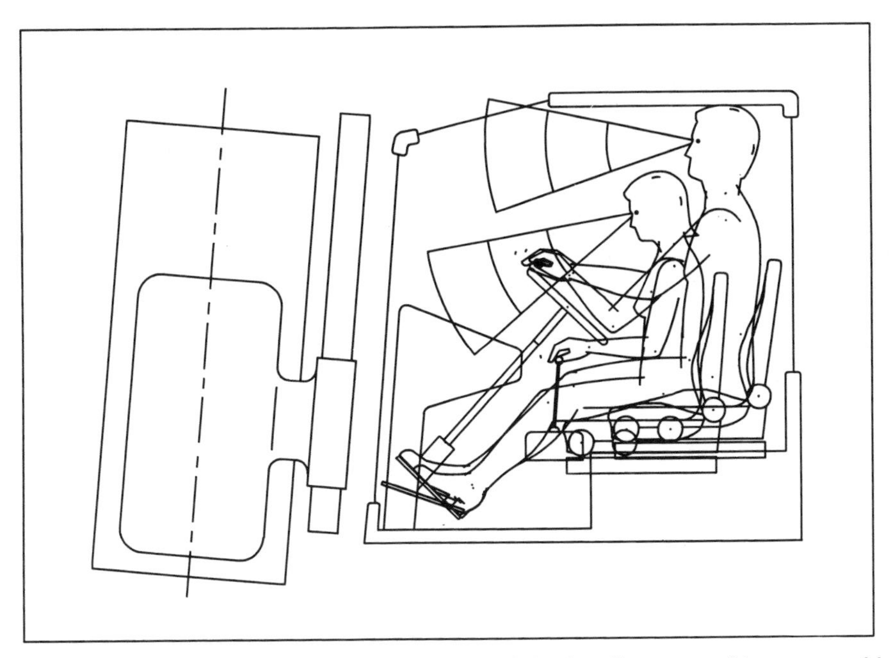

Figure 15.3.Assessment of a low harbour truck with the help of small woman- and large man-model.

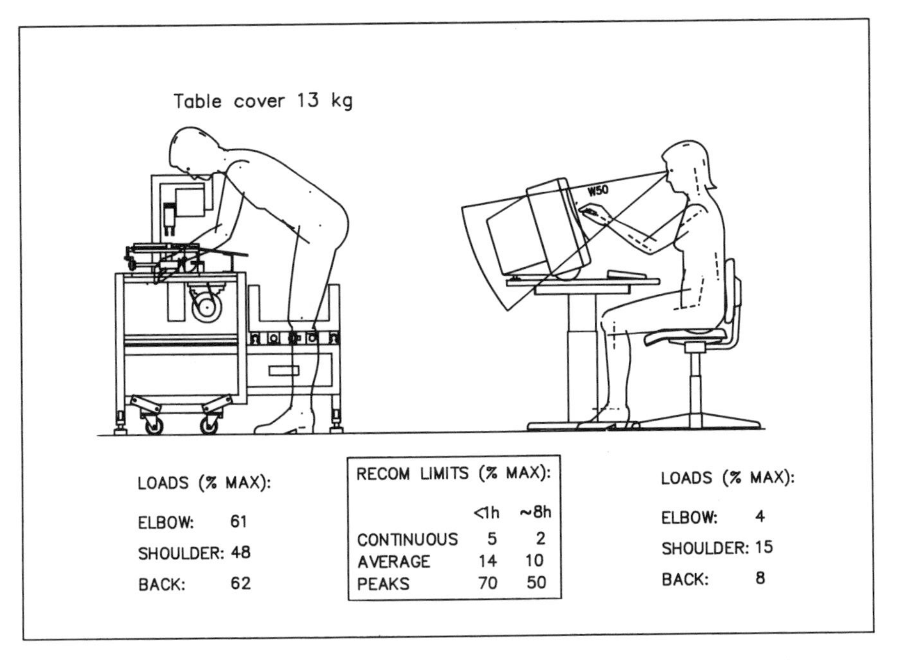

Figure 15.4. Biomechanical output of two different working situations with suggested recommendation values.

The anthropometric data are based on Finnish, North European and North American populations (see, for example, Heliövaara and Aromaa, 1980; DIN, 1986; Webb Associates, 1978). However, the man-models can easily be linearly scaled to other populations.

The evaluation of postural stress is based on mechanical modelling of the human body (see, for example, Chaffin and Andersson, 1984; Jonsson, 1984). In the actual posture with the determined external load, the biomechanical program calculates the stress as a percentage of the maximal static muscle strength (Figure 15.4). The percentage values can be referred to the recommendations displayed for various situations or durations. The biomechanical calculation at present takes into account only vertical loads. The individual variation can be taken into account and a wide scale of exertion levels can be assessed–from light static loading to a momentary high exertion.

The ergoSHAPE system is operated inside the CAD program and can therefore easily be expanded by either institutions or single users of the system. Special recommendations, such as layout and dimensioning charts for different purposes, can be added to the system. In this respect the system differs considerably from earlier man-model programs, to which modifications cannot generally be made by the user.

Study of use

Aim of the study

In order to evaluate the requirements for use of the ergoSHAPE system, a study was carried out in three enterprises. The aim of the study was to investigate:

- the present design practice of the workplaces (e.g. how the plans are documented and how ergonomics is taken into consideration) and the role of different occupational groups in the design process
- how the different types of man-models are suited for different design tasks
- how the different occupational groups are able to use the man-models and to assess the plans of the workplaces with them
- how strong a reference the man-model is for dimensioning a workplace
- how the assessment of the man-model corresponds to that of an actual human being.

One of the most important questions was the applicability of two- versus three-dimensionality to the designing of workplaces. For the sake of comparison, the three-dimensional man-model program called ergoSPACE was developed to comply with the restrictions of microcomputers. The man-models can be presented in stick and ellipsoidal forms. The environment is structured as simplified wire models, i.e. as lines, squares and boxes. Hidden lines cannot be removed (Figure 15.5).

Material and methods

The study was carried out in three different metal-product industries, where the so-called JIT (Just-In-Time) production principle had been adopted. All the occupational groups which influence the designing of the workplace in one way or another were represented in the study, including industrial engineers, supervisors, occupational health and safety personnel and workers.

The following methods were applied:

- interviews of all subjects about the present design practice
- assessment of the technical drawings. All subjects were asked to generate improvements to the plans and to evaluate the usefulness of the models for this

improvements to the plans and to evaluate the usefulness of the models for this purpose. The technical drawings were supplied with different kinds of man-model and recommendation charts

- design tasks with the CAD program. The industrial engineers applied the different man-models and recommendation charts to their own specific design problem. The subjects were asked to evaluate the usefulness of the models
- experiments to determine the optimal and acceptable workplace dimensions for a particular task with a man-model. All subjects assessed the comparative series of drawings and photographs, where the man-model and an actual man were set to different postures required by a lifting task. The working heights were widely varied in the series of drawings and photographs. In addition, an optimal working height and an optimal working posture for a packaging work task were determined with a plastic six-part manikin drawn by ergoSHAPE.

Results

The interviews revealed that the work task and workplace were not generally well perceived in the design phase. The workplace plans were poorly documented, being restricted to occasional layout sketches. Detailed plans of a workplace were rare, normally the only documentation being on a space reserved in a general layout plan. The workplace was a result of several decisions and detailed plans concerning production technology and material handling, and the arrangements of the work and workplace were left mainly to the supervisors.

In the case of designing a highly mechanised workplace, the documentation was complete (including side view projection), but the physical human work was not seen as an important or critical design factor. The occupational health and safety personnel did not participate routinely in the design phase, except in problematic cases. The man-models were of great help to the occupational health and safety personnel in assessing the technical drawings. The engineers and the supervisors were able to find rough ergonomic deficiencies in the drawings without using man-models, which were useful mainly for the fine adjustment of the workplace dimensions.

The preferences concerning the different kinds of man-models and different projections were quite similar among the industrial engineers and other occupational groups. In two-dimensional top view projection of the ergoSHAPE system, the recommendation charts were preferred to the man-models, whereas in side view projection the nine-part models (including the model for biomechanical calculations) were preferred to more simple models. Both top and side views were considered necessary in ergonomic assessments.

In the three-dimensional ergoSPACE system, ellipsoidal presentation was preferred to stick-model presentation. The majority of the subjects found the three-dimensional design confusing, but some of them considered it the best way to get an overall picture of the work situation. Generally, the detailed two-dimensional plans were preferred for accurate ergonomic assessments.

The results of the design tasks supported the above observations. The dichotomy of preferences was strengthened. The schematic recommendation charts were accepted favourably and the most complicated and thus most natural nine-part model and biomechanical calculation aroused great interest. The three-dimensional system was not considered suitable for design purposes in its present form.

In the determination of optimal and acceptable dimensions for the workplace using the man-model, only slight differences were found between the occupational groups. The

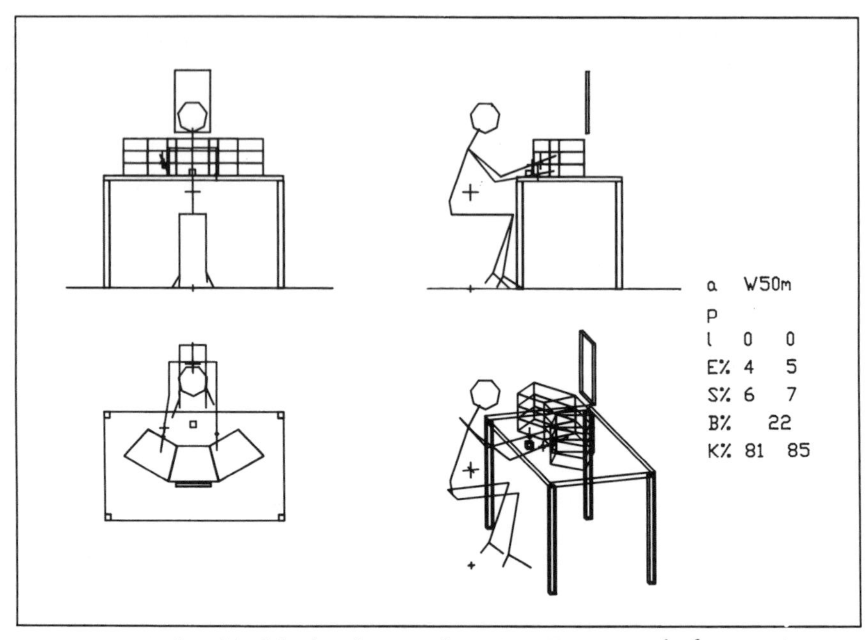

Figure 15.5(a). Stick-model of the three-dimensional ergoSPACE system in the four-projection presentation.

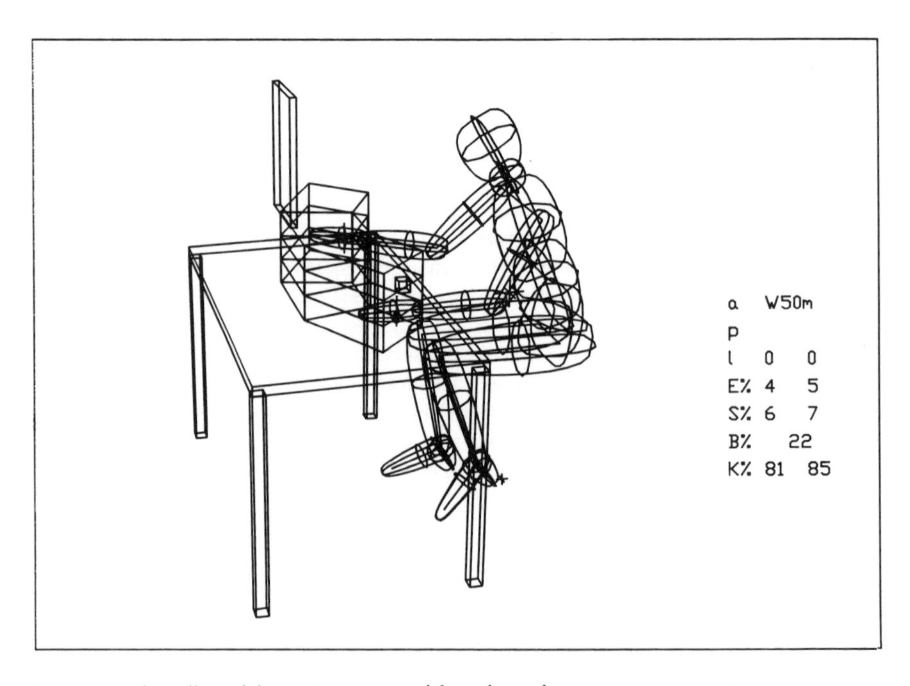

Figure 15.5(b). Ellipsoidal ergoSPACE model in the real perspective presentation.

differences in the determinations within the groups were more obvious. Ideally, according to the task description, the results should be the same as regards the different series of pictures. In the lifting task, the results of the industrial engineers were the most even in all respects, whereas the results of the occupational health personnel were the most uneven. In the packaging task, the variation in the determinations was considerably smaller.

Conclusions

There was considerable discrepancy between the actual design documentation of the workplaces and the ergonomic man-model presentation in the three enterprises tested. The design practice concentrated mainly on horizontal action (material flow), whereas the main interest from the ergonomic point of view lies in the vertical dimensioning of the workplace, which usually has a greater effect on the physical load on the worker. This implies that the planning must be at a more advanced stage before the man-models can be utilized.

Because industrial engineers are generally not very interested in the detailed designing of workplaces, it is obvious that the main area of application of the man-model systems is still limited to the designing of workplace products, e.g. large machines, cabins and furniture. On the other hand, there is a trend towards using ready-made components and construction entities. Workplace design in enterprises already consists to a great extent, and will consist even more in future, of buying and installing machines and equipment without detailed technical drawings.

The designing of work and workplaces is holistic by nature, and therefore any separate ergonomic system does not appeal to the planners. The integration of ergonomic knowledge with the complete design data, within the CAD system is therefore essential in the future.

The fitting of technical elements to each other and to the worker is the most complicated phase in designing, and it takes place in one single head, that of an industrial engineer. Both expert cooperation with, for example, occupational health personnel, and user participation are natural before and after this phase. The man-models in workplace documents are perceived as very useful in facilitating the feedback to the planner.

The man-model is a very strong reference for workplace measurements, leading to a similar result within different occupational groups. Basic education and general experience have only slight influence on the ability to assess plans ergonomically and to determine dimensions with the man-model. Familiarity with the actual working situation and the work process is far more important.

The natural appearance of the man-models tends to be more important than small differences in the ease of their technical use. In order to cover all design situations, versatile and naturalistic man-models are needed together with the schematic recommendation charts in the ergonomic design system, and preferably in a from in which they can easily be combined.

The simplified three-dimensional system is obviously not useful for designing workplaces, partly because it is not connected to existing ways of documenting the workplaces, and partly because of its over simplified, unnatural and confusing visual appearance. A more detailed modelling of the environment would probably require too much design effort, especially if the design problem is not expected to be critical, and the three-dimensional designing is not felt to be necessary.

The biomechanical calculations have been interesting to the engineers as well as other occupational groups. Although the use of human models releases designers from following rigid recommendations, it presumes a greater understanding of human functions. The biomechanical calculations can be considered as a means of teaching the planner a more ergonomic way of thinking, and of encouraging him to include more consideration of the human being in the design stage.

References

Chaffin, D.B. and Andersson, G. (1984). *Occupational Biomechanics*. (New York: John Wiley).

DIN (1986). *DIN 33 402, Teil 2: Körpermasse des Menschen*. (Berlin: Beuth Verlag GmbH).

Heliövaara, M. and Aromaa, A. (1980). Suomalaisten aikuisten pituus, paino ja lihavuus. *Kansaneläkelaitoksen julkaisuja ML: 19*, Helsinki (in Finnish).

Jonsson, B. (1984). Rörelseorganens funktionella anatomi och biomekanik. *Arbetarskyddsstyrelsen, Utbildning* **12** (in Swedish).

Launis, M., Lehtelä, J. and Kuusisto, A. (1988). Kaksiulotteinen antropometrinen ihmismallijärjestelmä työpaikkojen tietokoneavusteiseen suunnitteluun (CAD). *Työ ja ihminen*, **1**, 68–83 (in Finnish).

Webb Associates (1978). *Anthropometric Source Book, Volume I, NASA Reference Publication 1024*. National Aeronautics and Space Administration, Scientific and Technical Information Office, Washington D.C.

Chapter 16
Applications of the SAMMIE CAD system in workplace design

K. Case, M.C. Bonney, J.M. Porter and M.T. Freer
SAMMIE C.A.D. Limited, Quorn, Loughborough, UK

Introduction

Computer Aided Design (CAD) is now firmly established in some industries as the normal method of originating and evaluating designs. Thus in aerospace it would be normal to have computer representations of proposed aircraft long before mock-ups or prototypes are available for functional evaluation. This implies that many aspects of the design may be finalised before there is any opportunity to carry out ergonomics evaluations of the work space or work tasks which will eventually confront the operator. Other industries are not so advanced in using computers in design, but would benefit from the ability to carry out ergonomics evaluations early in the design process. It is natural therefore to look for CAD systems which have the capability of considering human as well as mechanical, structural or other aspects of design.

SAMMIE, System for Aiding Man-Machine Interaction Evaluation, is one such system which has been used in this way for some years. It assists in the building of a computer model of the workplace which can be viewed and manipulated on a graphics screen in ways which will be familiar to users of modern three-dimensional solid modelling systems. In addition, and most importantly, it includes a model of the human operator which is used as an evaluative tool.

This paper very briefly describes the characteristics of SAMMIE but concentrates on describing applications of the technique to workplace design. In the main these applications originate from design consultancy carried out in recent years, and include supermarket checkout facilities, visibility studies in underground trains, and a machine shop environment.

The SAMMIE System

Details of the SAMMIE system can be found elsewhere (Case, Porter and Bonney, 1986 and 1989) and only a brief description is given here. Workplace modelling facilities are provided by a relatively simple form of boundary representation solid modelling (Requicha, 1970). Solid shapes are represented by plane-faceted polyhedra which are usually specified by the user as primitive shapes such as cuboids, cylinders, cones, prisms, etc. These are combined with spatial and relational information into a hierarchical data structure which permits the modelling of certain functional aspects (such as the opening of a car door). Interaction with the model is via a menu system and most of the normal CAD system facilities are available. Thus the model may be displayed in various ways (with hidden lines removed, colour or monochrome, etc.) and viewed as appropriate (i.e. in plane parallel projection, orthographic projection or perspective). The model may be edited to change dimensions of objects, reposition or reorientate them, and to change functional relationships (such as changing ownership within the data structure).

A man modelling facility is based upon and completely integrated with this workplace modelling system. A pin-joint and rigid link approximation is used to define the model's structure and this is encased in solid modules of 'flesh', (as shown in Figure 16.1). All aspects of this are data-driven and most are accessible to the user, thus ensuring that the model represents the desired user population. The joint-to-joint dimensions (or limb lengths) can be specified in a variety of ways including the selection of percentiles from population data. Similarly the flesh shape is controlled by somatotype specification (Sheldon, 1940). Articulation about the joints is possible so that working postures can be created and evaluated. Commonly used postures (for example standing and sitting) can be directly called from a menu, while others can be retrieved from previously created databases which may be specific to an application area. Alternatively, the model may be moved interactively a joint at a time, or reach algorithms may be employed to predict postures related to a task. In all cases joint extension is constrained to remain within maximum limits which once again are defined by user-accessible data. It is also possible to evaluate postures against

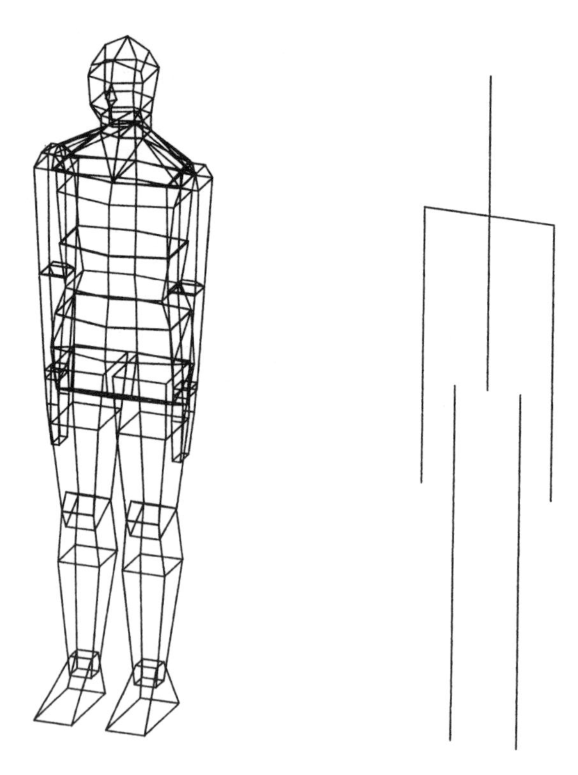

Figure 16.1. The man model showing the 'stick' and 'flesh' representations

'normal' constraints which could for example represent desirable working postures. Special purpose facilities are provided to allow the assessment of reach within areas or volumes of the workspace. An envelope of reach (maximum or normal) can be predicted for any surface of the model dependent on the anthropometry of the man model. Similarly, volumes of reach can be predicted (Figure 16.2). Extensive vision assessment facilities are provided over and above the normal model viewing techniques described earlier. The view as seen by the man model can be displayed, permitting a visual field assessment which is particularly useful in cockpit situations where visibility is potentially obstructed by the bodywork of the vehicle. A three-dimensional vision chart (Porter, Case and Bonney, 1982) has been developed to give a characteristic of all around visibility. For similar purposes, Aitoff projections (Figure 16.3) can be produced to assist in compliance with MIL standards (MIL850B,1970).

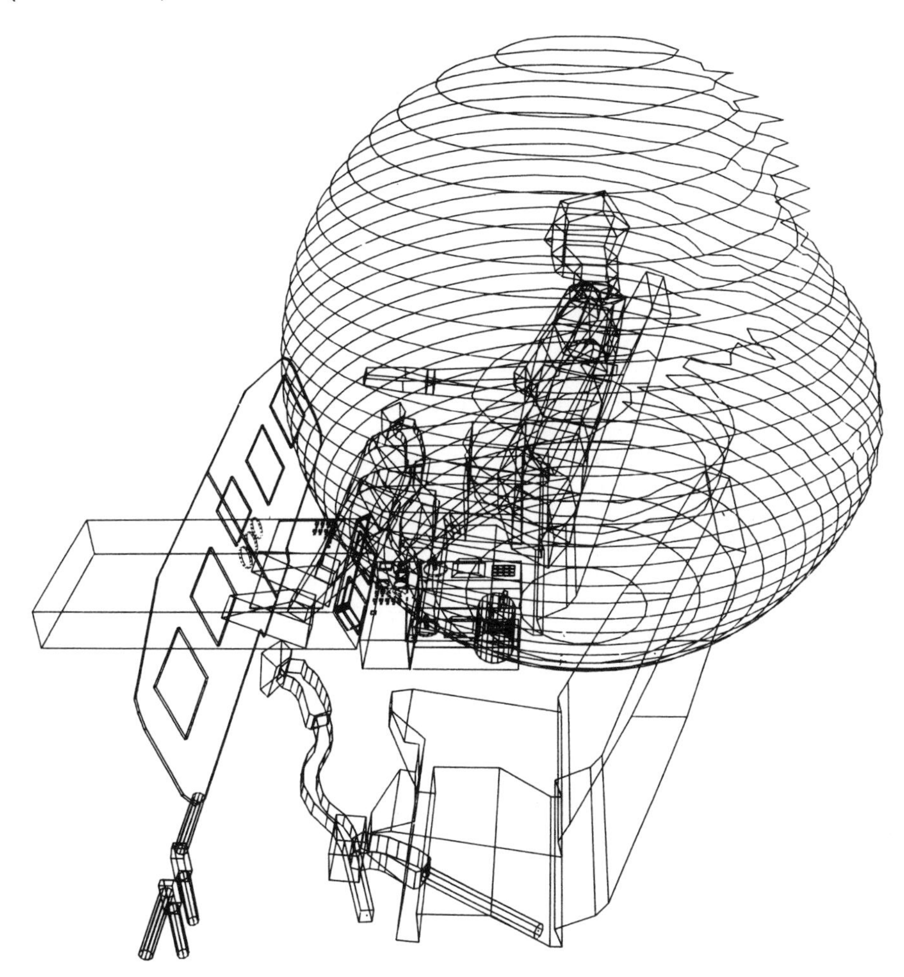

Figure 16.2. Use of the volumetric reach facility within a helicopter cockpit

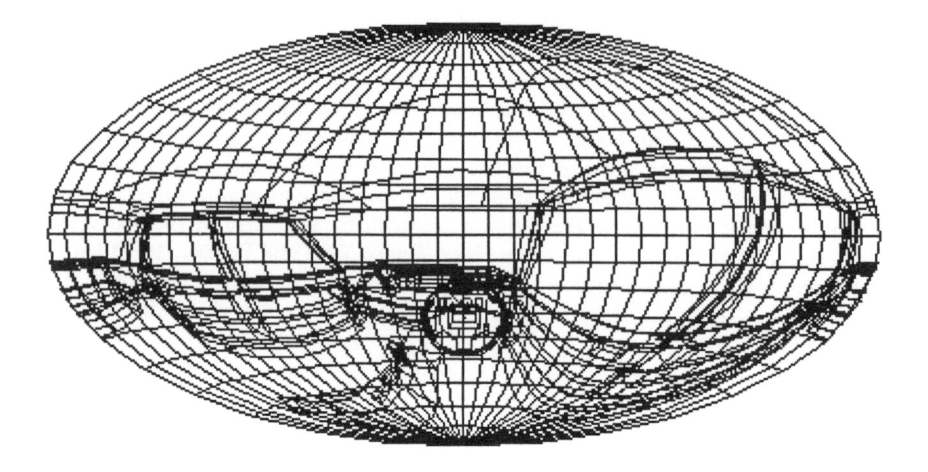

Figure 16.3. Typical Aitoff projection showing 360 degree visibility from a car

Applications

Reach zones for agricultural tractors

The driving of agricultural tractors presents several areas of difficulty which would not normally be found with more conventional vehicles. The working environment is likely to be noisy and rough terrain can lead to an uncomfortable ride, a problem which may be exacerbated by extended driving periods perhaps in extremes of heat or cold. The controls are likely to be more complex than a road vehicle, since in addition to the driving task, the driver would be expected to operate a variety of agricultural implements. Alleviation of these difficulties may be tackled in several ways and recent tractor designs have shown a marked improvement in the standard of environmental comfort for the driver. In terms of easing the work task, improvements could be made if controls were located for convenience of reach rather than ease of engineering design. SAMMIE was used to investigate these problems at the feasibility stage of tractor design (Porter, 1979), and a need to determine suitable reach zones was identified (see Figure 16.4). Further work (Reid *et al.*, 1985) provided a general framework for defining zones of reach as volumetric envelopes for males and females of various percentiles. More importantly, these zones related to working postures rather than to artificial laboratory conditions. Using SAMMIE, extensive sets of data were generated which later formed the basis of a British Standard (BS6375, 1987).

Supermarket checkout facilities

Supermarket checkouts provide a typical example of the use of SAMMIE to compare existing workplace designs with proposed alternatives. Models, as in Figure 16.5, were built of the old and new designs and the man model was specified to reflect the anthropometry of the expected user population. In this case anthropometry was based upon British adult females aged 19 to 65 years and was obtained from Pheasant (1984). This was deemed to represent the major users of the equipment although some brief use was also made of

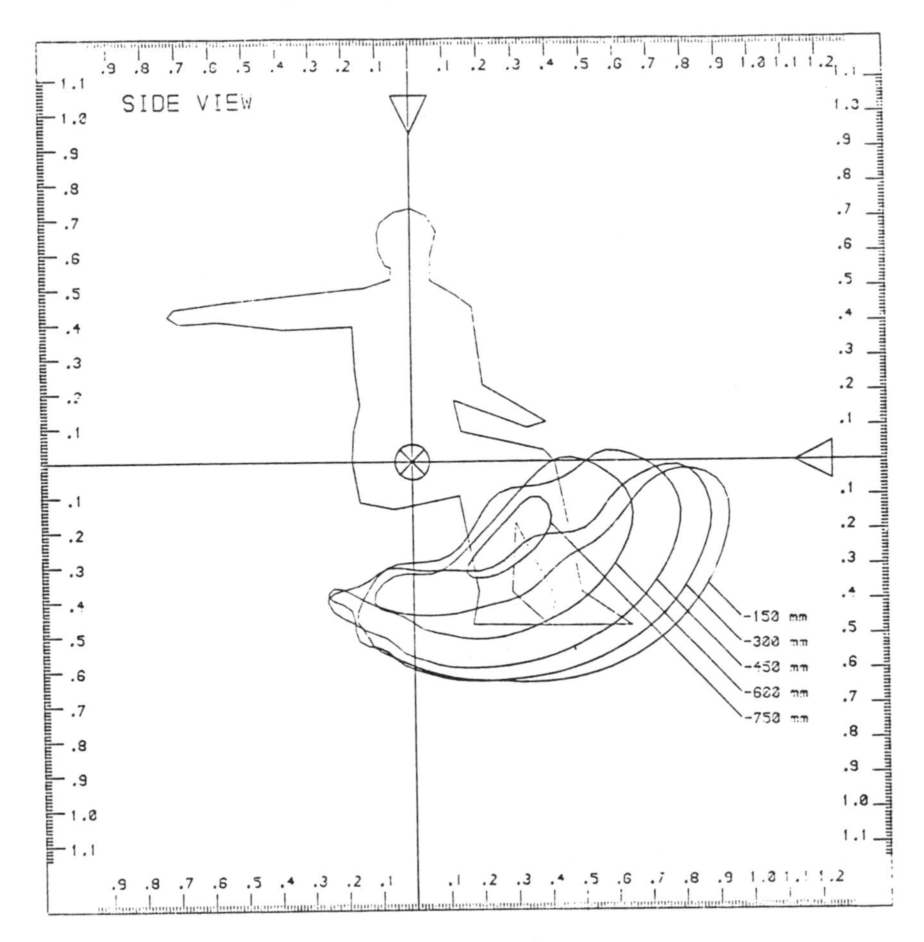

Figure 16.4. Example of a reach zone for agricultural tractor drivers

male anthropometry. The main criteria to be evaluated related to fit, reach and working postures whilst performing a variety of tasks from a standing or sitting posture. Thus 5th percentile females, 95th percentile males of both average and fat build were used to assess leg room, thigh clearance, seat height and foot support. Eye contact between the checkout operator and the customer was considered important and it was possible to evaluate this for a variety of counter heights, anthropometry, and sitting or standing operation. Other aspects of the work task evaluated included the determination of reach zones both maximum and comfortable (working limits for reach without stretching) for the specified user population. The layout of equipment such as the printer, scales, keyboard, or cash drawer was analysed with respect to reach, task sequence, functional grouping, and importance and frequency of use. The value of three-dimensional modelling of both workplace and operator, as opposed to just two-dimensional design of the workplace, is evident from the design recommendations made. In many cases these related to postural difficulties arising from the confined nature of the workplace, and the somewhat awkward postures required for some tasks. Thus for example it was shown that one design constrained leg and knee room to such an extent that some swivelling movements required to reach items of equipment were not possible for some of the large members of the user population. Traditional design methods would have been unlikely to discover these aspects until user trials were carried out on prototypes.

Figure 16.5. Design of supermarket checkout facilities

One-person underground train operation

SAMMIE is particularly useful for visibility studies, especially where the operator adopts unusual working postures. In the case of one-person operation of underground trains, the control of the automatic opening doors is in the hands of the driver. For safety reasons he would normally be required to leave his seat at each station and to look back down the platform to ensure that there would be no problem in closing the doors. With the heavy utilisation of the London Underground system, the extra delay as the driver returns to his driving position is to be avoided if at all possible.

An alternative working method is for the driver to remain in his seat and to view the situation on the platform via mirrors or video monitors strategically placed on the platform forward of the stopping position of the train. In this case the driver must be able to view the mirrors or monitors through the front and side windscreens. The viewing problem is further complicated by the necessity to use windscreen wipers which only sweep part of the screen surface. An earlier study had established the sizes and locations of the mirrors and video screens on the underground line in question, and these had been formed into envelopes representing the maximum viewing requirement on both the nearside and offside. Driver anthropometry was represented by 5th percentile female and 95th percentile male models which were permitted to adopt four working postures—sitting upright, leaning forwards, leaning forwards to the left and leaning forwards to the right. These postures were considered acceptable given the intermittent nature of the task. A general view of the situation is shown in Figure 16.6. Figure 16.7 shows one of a number of views as seen by the driver from one of the working postures. The study established that the proposed one-person operation was feasible for this aspect of operating the automatic doors from the seated driving position.

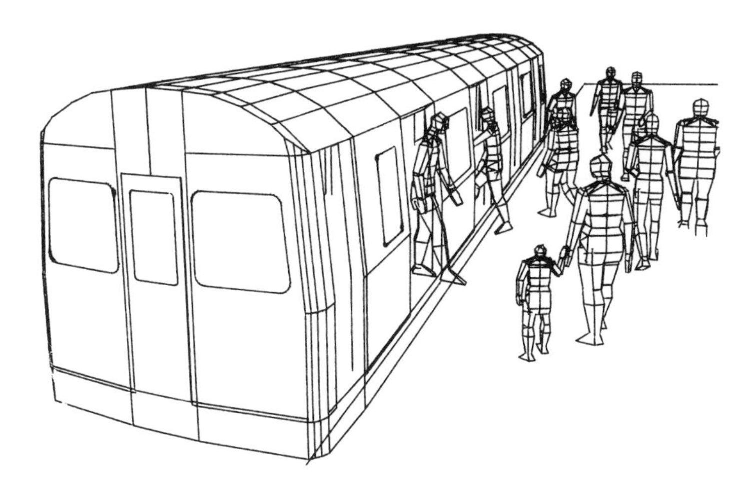

Figure 16.6. General view of passengers boarding underground train

Machine shop environment

Figure 16.8 shows part of a model of a machine shop which has been used to evaluate maintenance procedures for large and complex numerically controlled machine tools such as the Wadkin 3-axis, automatic tool-changing machining centre. The Coordinate Measuring Machine (CMM) which is also shown in the figure has recently been added to the model with the intention of studying the working postures that must be adopted to operate finely controlled equipment.

References

British Standards Institution (1987). *Reach volumes for location of controls on agricultural tractors and machinery, BS6375.*

Case, K., Porter, J.M. and Bonney, M.C. (1986). SAMMIE: A computer aided design tool for ergonomics. In: *Proceedings of the Human Factors Society Annual Conference, Dayton, Ohio, USA.*

Case, K., Porter, J.M. and Bonney, M.C. (1990). SAMMIE: A man and workplace modelling system. In: *Computer aided ergonomics*, edited by W. Karwowski, A. Genaidy and S.S. Asfour. (London: Taylor and Francis).

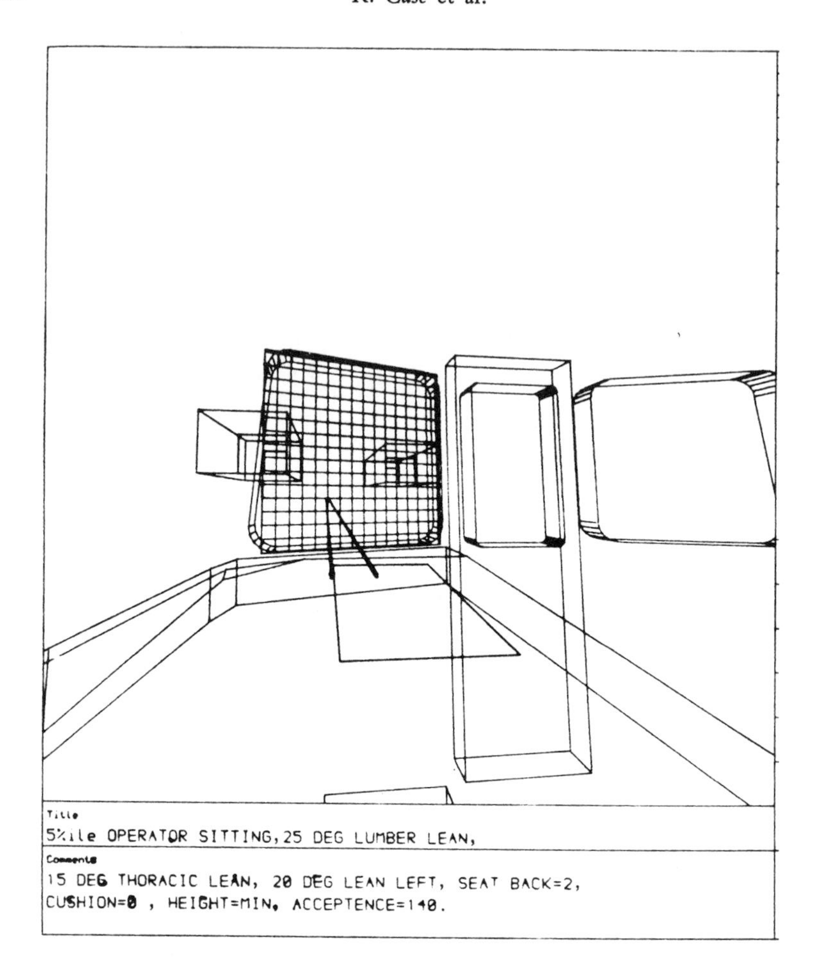

Figure 16.7. Driver's view of mirrors and monitors from across the cab. View shown for the 5th percentile female leaning forward and towards the left

MIL (1970). *Aircrew Station Vision Requirements for Military Aircraft*, MIL Standard 850B.

Pheasant, S. (1984). *Bodyspace: Anthropometry, Ergonomics and Design*, (London: Taylor and Francis).

Porter, J.M. (1979). Control Design and Layout for a Tractor Cab. Report No CAS.018, Department of Production Engineering and Production Management, University of Nottingham.

Porter, J.M., Case, K. and Bonney, M.C. (1982). A computer generated three dimensional visibility chart. In: *Human Factors in Transport Research*, edited by D.J. Oborne and J.A. Levis, (Academic Press) Vol. 1, 365–374.

Reid, C.J., Bottoms, D., Gibson, S.A. and Bonney, M.C. (1985). Computer simulation of reach zones for the agricultural driver. In: *Proceedings of the Ninth Congress of the International Ergonomics Association*, September 1985, Bournemouth (London: Taylor and Francis)

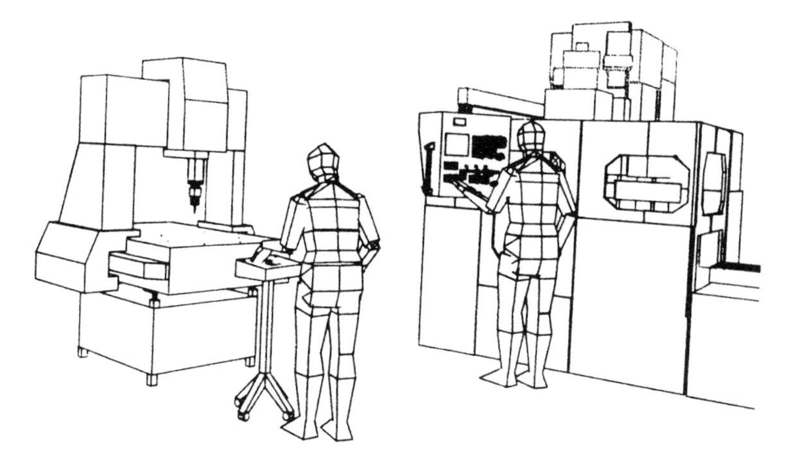

Figure 16.8. Machine shop showing a Wadkin Machining Centre and a Co-ordinate Measuring Machine (CMM)

Requicha, A.A.G. (1970). Representations for rigid solids: theory, methods and systems. In: *Computing Surveys, 12.*

Sheldon, W.H. (1940). *The Varieties of Human Physique*. (New York: Harper Brothers).

Part V

Case studies: ergonomic improvements through workplace design

Five studies present workplaces from widely diverse industries and two continents. The working situations range from the harsh environment of fishing in the North Sea to the pressures of repetitive precision tasks at the assembly bench. Such practical examples, discussing the factors influencing work, can be valuable in seeing the interplay between organisation and working conditions.

This is highlighted in Stoop's investigation of the design of Dutch fishing vessels, which gives a graphic account of the tasks involved. Moreover, Stoop proceeds to outline a technique for analysing the potential for human error, based upon an application of Reason's Generic Error Modelling System, and applies this to the design of sophisticated equipment used on the bridge.

Nygård and Ilmarinen provide a contrast in their study of delivery drivers in Finland, looking at the manual handling of dairy products. Here, the workload is mainly physical, and can be much improved by consideration of the packing methods and of the equipment which is available. However, time spent on different activities and resulting physical strain of the job can equally be affected by the method of transport and site facilities, so that the effects have to be considered throughout the whole delivery chain.

The next three papers all come from manufacturing industry, where major problems arise from the design of the layout of workstations and working methods. Bhasin, Kothiyal and De studied the layout of press operations, Dellemann and Dul the design of sewing machines, and Kothiyal, Bhasin and De the workplaces used for electronic assembly. Here, postural strain is a major concern, and can be reduced but not necessarily eliminated by ergonomic layout of the workplace. The work is highly repetitive and may be paced, and other factors such as motivation need to be considered. As all the case studies here demonstrate, a variety of ergonomic techniques are needed to assess working situations—it is not sufficient to consider only one aspect of the working task.

Chapter 17
Redesign of bridge layout and equipment for fishing vessels

J. Stoop
Safety Science Group, Technical University Delft, The Netherlands

Introduction

After the Second World War, major changes occurred in the sea fishing industry in The Netherlands. During a period of economic growth and technical improvements, beam trawling became the major fishing method for small to medium sized fishing vessels. An enormous increase in catch capacity occurred through a scaling up of propulsion power and fishing gear size and by the application of electronics and information technology. Today, the beam trawler fleet comprises 600 ships, with a total power capacity of 538,000 hp and employment for approximately 3,000 men.

Data supplied by the Dutch Shipping Council show an increase in the involvement of fishing vessels in collisions in the North Sea during the last decade (Table 17.1). Besides these collisions there has been an increase in the number of 'near misses' (van der Sloot, 1987).

Table 17.1. Numbers of vessels involved in collisions in the North Sea

Year	1978	1979	1980	1981	1982	1983	1984	1985	1986	1987
Total	23	26	28	20	15	30	32	39	29	19
Fishing	9	9	10	10	9	12	22	23	20	9

Accident analysis

Collisions

The increase has been caused by a number of background factors. The intensity of North Sea traffic has increased, the introduction of deep water routes and traffic separation lanes have concentrated the traffic in shipping lanes, which are hazardous to cross especially during the fishing process. Some fishermen even want to fish in and between these lanes. The limitation of the allowed fishing days by the European Community policy has caused production stress during fishing to catch the individual allowance of fish. The increase in undermanned coasters with frequently unmanned bridges introduces a new category of hazardous vessels to the North Sea area.

Analysis of the Shipping Council accident data gives various indications (van der Sloot, 1987). In a number of cases, ships were unaware of the presence of other ships because of a low expectation of other traffic, pressure of work or fatigue after a week of continuous fishing.

Wrong interpretation of the radar has occurred many times, even by experienced watchmen, because it is difficult to convert the distance and course changes on the screen into a usable cognitive model of reality. Radar interpretation is often hampered by rough seas, and cloud clutter and resolution problems with daylight screens. Because of reduced manning, fishing vessels also rely on unqualified and untrained young watchmen. These watchmen may also suffer from fatigue problems by staying out late the night before departure and because of the use of alcohol.

Accidents occur as frequently by day as by night, during good and poor visibility. Poor visual conditions are often combined with incorrect radar interpretation, while preoccupation with additional activities may occur in good visibility. The heavy work load with a minimum crew number forces the watchmen to perform additional activities which are not included in the actual bridge tasks, so that the bridge tasks are easily neglected. It must be pointed out however, that statements from watchmen to the Shipping Council that they were involved in additional activities at the time of the accident are not necessarily true, but may be 'acceptable' excuses for the causes. Collisions with overtaking vessels are often caused by poor rearward visibility from the bridge.

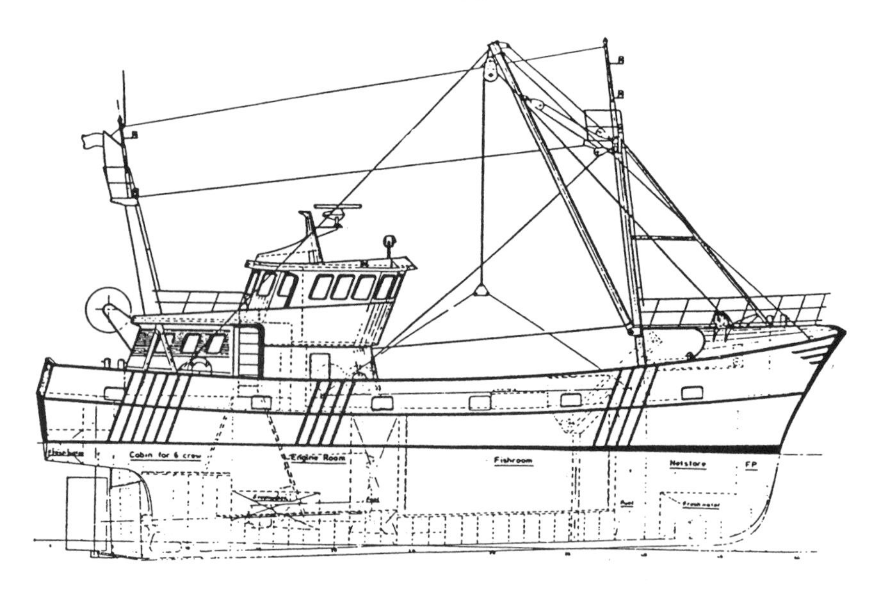

Figure 17.1. Beam trawler

Capsize and sinking

Between 1981 and 1988 five ships capsized and sank due to overturning the port or starboard outrigger, positioning the fishing lines over the wrong side of the ship or by unequal pulling force on the fishing lines during washing of the nets or by jamming of the nets against sea bottom obstacles such as wrecks or stones.

A case study

A beam trawler capsized and sank at night while fishing (Raad voor de Scheepvaart, 1988). It is a normal procedure during fishing to wash the nets when they collect too much sand or sea weed. The watchman can do this task by himself and needs no assistance. He reduces power, hauls in the fishing lines, slightly tops the outriggers and gains headway until the nets are clean. To continue the fishing process he reverts to the former position. The watchman involved in the accident reduced power, assumed he had put the handles of both the fishing lines in the 'haul' position and left the bridge to look after his soup which was heating, while the winch was still running. The controls of the fishing lines and the outriggers are positioned close to each other on the front console and are identical. On his return, he switched on the deck lights, noticed the starboard outrigger in a vertical position against the mast and realized his mistake. He unsuccessfully tried to lower the outrigger, but failed to let the starboard fishing line run out and failed to stop the engine. The ship heeled heavily to port, whereupon he warned the crew without taking further actions. The ship heeled further to port and rapidly flooded through the open engine room door at the port side and through the open fishing hold. All crew members were saved.

Occupational accidents

In a number of cases the man on the bridge was involved in accidents which occurred on the deck. These occurred during the hauling in of the nets while manoeuvring the ship, or during repair and other activities while operating the winches of outrigger and fishing lines from the bridge (Raad voor de Scheepvaart, 1984).

A case study

During fishing the net frequently tears because of an overload of mud and sand and must be retrieved. The crew members on the deck were experienced and performed this task without the need for instructions. The crew was well aware of the hazards because of previous accidents on other ships. During the turn to port to retrieve the net, the skipper oversteered and a line caught on the starboard corner of the stern. The skipper was not using the starboard steering console, but the central steering position which deprived him of any sight of the aft deck. Suddenly, the tightly stretched line came loose with great force from the stern and injured the three crew members on the aft deck. The skipper was concentrating on the torn net and only noticed the accident when one of the crew members walked into his sight. The crew members were treated for their injuries.

Analysis of the accidents indicates that there are clear problems in carrying out the bridge tasks under current conditions, and that there is a need for further study.

Analysis of the bridge tasks

Navigation

During the last decade much attention has been paid to bridge design for deep sea merchant vessels. Rationalisation and automation has led to the 'Bridge 90' concept: a one-watchman automated bridge (Schuffel and Boer, 1985). A careful allocation of tasks in the man-machine system has resulted in an automatic bridge fit for safe and smooth performance of executive and supervisory tasks. Research on the mental load in three different bridge configurations has shown that:

- the accuracy of the navigation task is at its best in the Bridge 90 concept. A conventional bridge with two men scored lower accuracy and a one-man conventional bridge the lowest accuracy
- errors in a continuous memory task did not differ significantly between a control condition, the Bridge 90 concept and the two-man conventional bridge, but the error score on the one-man conventional bridge suggested a mental overload.

Analysis of 100 Dutch shipping accidents showed that 68% of the 209 events occurring during the accidents could have been prevented by use of a Bridge 90 concept (Schuffel, 1988).

The fishing process

For the fishing industry a supplementary task is added to those found on merchant vessels: the fishing process. The work on board a fishing vessel is heavy, with continual labour from Monday to Friday in a 2-hour working cycle (repeated 40 times) of emptying the nets on deck and processing the catch. This is added to high noise and vibration exposure, a combination of living and working in a moving workplace, with climate and weather conditions and stress factors due to catch quotas and governmental inspections (de Regt, 1987; Veenstra, 1986).

Field observation of behaviour on the bridge of a beam trawler indicates a great fluctuation in work load, with periods of both considerable over- and underloads (Heinrich, 1988a). Overloads occur during the shooting and hauling of the nets; an underload occurs during the outward and homeward voyage. In contrast to expectations, no significant difference in load was found between the periods before and after the fishing as far as the navigational task was concerned. During the fishing process a high average load was found, with a difference between the trawl itself and shooting or hauling of the nets. The trackplotter and echo-sounder were scrutinized by the skipper continuously during the trawl to keep the desired track and to prevent the nets from jamming. The winch operation demanded full attention during shooting and hauling of nets. While operating the winch the skipper leaned out of the window to judge the catch, for control purposes and to give instructions to the crew. This allowed very little attention for navigation. During the outward and homeward voyage the watchman had nothing to do but to supervise the radar and trackplotter.

Conclusion

The mental load on a one-man conventional bridge is heavy compared to a two-manned conventional or one-man automated bridge. A supplementary work load is added by the fishing process, particularly during shooting and hauling of the nets, leaving little room for navigational tasks.

Problem development

A normative task description

To picture the activities on the bridge, a normative task description was developed, which comprises all subtasks in a time sequence (Draijer, 1988). The normative task distinguishes the different phases of the journey and is represented schematically as follows (Heinrich, 1988a):

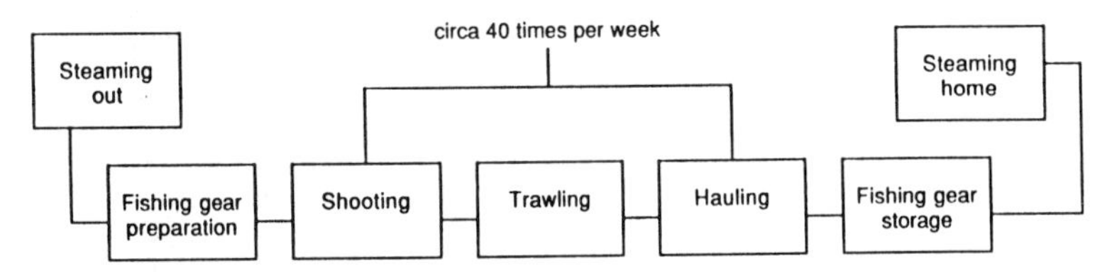

Figure 17.2. Normative task description of bridge activities

In the execution of the tasks we can distinguish a number of subtasks:

- navigation; preparation and execution
- steering; setting of the course and supervising the course changes
- communication with other ships and shore
- propulsion; supervision and control
- preparation and control of auxiliary power supply
- control and handling of support systems, fish hold and ship conditions
- supervision of the deck activities
- detection and pursuit of flat fish

Instrumentation

The following instruments are available for navigation:

- magnetic compass to set the course and bearing of other ships, which may be linked to the autopilot
- gyroscopic compass, more accurate than a magnetic compass but far more expensive
- navigation systems to locate the position of the ship, especially by decca and radio beacons. In the future satellites will be available, linked to the trackplotter
- radar for information about surroundings
- trackplotter to indicate the track, wrecks and fishing grounds, linked to navigation instruments
- autopilot to maintain a preset course, linked to the compass. The autopilot is linkable to the trackplotter and radar in order to maintain a preset track
- communication systems such as mariphone, radio-telephone, satellite communication, navtex installation
- Doppler-log to indicate the vessel speed over the sea-bed
- ARPA (Automatic Radar Plotting Aid) which will become available in the future to indicate the speed and direction of targets visible on the radar and equipped with an adjustable 'guard' zone. GPS (Global Positioning System) satellite navigation will also become available
- watchkeeping alarm, which has recently been made compulsory to prevent the watchman falling asleep

For the fishing process the following instruments are available:

- echo sounder to picture the underwater situation directly beneath the ship, with respect to objects and the sea bottom and to determine the depth
- winch control equipment

There are different conditions of equipment use possible, which may add a supplementary load, such as weather and visibility conditions, traffic area, phase of the journey, time of the day or traffic situation. These conditions of use can be represented by different 'use scenarios'. The 'use scenarios' which probably contribute to an increase in work load or risk can be identified (Gelderblom, 1988; Stoop, 1987).

Scenario

	Fog	Storm	Traffic	Engine failure
Navigation	**		**	**
Steering	**	**	**	
Communication	**		**	**
Propulsion		**	**	
A.P.U.		**		
Support system		**		
Deck activity		**		

Figure 17.3. Use scenarios

Human error

The subtasks can then be analysed to identify errors and deviations which can induce risks (Kjellèn, 1984). In the pilot studies, attention was focused on human error through design limitation and shortcomings. The errors due to maintenance, malfunctioning in the equipment, electromagnetic interference between instruments or damage due to the corrosive environment or vibration are not dealt with here.

Errors can be described both in terms of effects and of causes. In the case of causes a detailed analysis is required. In the case of effects, the extent of detailing necessary is dependent on an evaluation of available solutions. If these solutions are proved to be inadequate, a further detailing into subproblems is required. Introduction of new solutions may also generate new errors.

There are two principles for relating tasks to errors. First we can learn by experience through analysis of accidents that have already occurred. Second we can apply human error theory to try to predict 'foreseeable' use and error, before accidents have occurred. Such theories should therefore have both an explanatory and a predictive potential.

Models of risk-taking behaviour on a aggregated as well as an individual level suffer from several shortcomings and have a limited applicability in this sort of problem (Janssen, 1986; Bos, 1988). Models known from cognitive psychology are much better able to link normal behaviour, human error and risk inducing factors which may lead to accidents (Hale *et al.*, 1988; Quist, 1987). Rasmussen (1980) distinguishes three levels of human functioning, which he calls skill-based, rule-based and knowledge-based, based respectively upon the carrying out of highly practised routines, the choice of appropriate courses of action and the development of new ways of coping with problems.

Reason (1987) bases his work upon the analysis of slips and mistakes, which he found to be characterised by the intrusion of 'strong-but-wrong' errors. These were highly practised programmes of action which captured control of behaviour, often at times when the person was preoccupied or distracted. Reason defines slips as departures of action from intention, or execution failures. Mistakes are errors in which the action may run according to plan, but where the plan is inadequate to achieve its desired outcome.

Wertheim (1977 and 1982) indicated a possible additional factor in inadequate reactions. When a person is operating with a low level of attention and an increase in attention is then required due to a sudden change in the situation, attention may not increase fast enough. This may occur in a situation where little or no variety exists over a long period in a supervisory task.

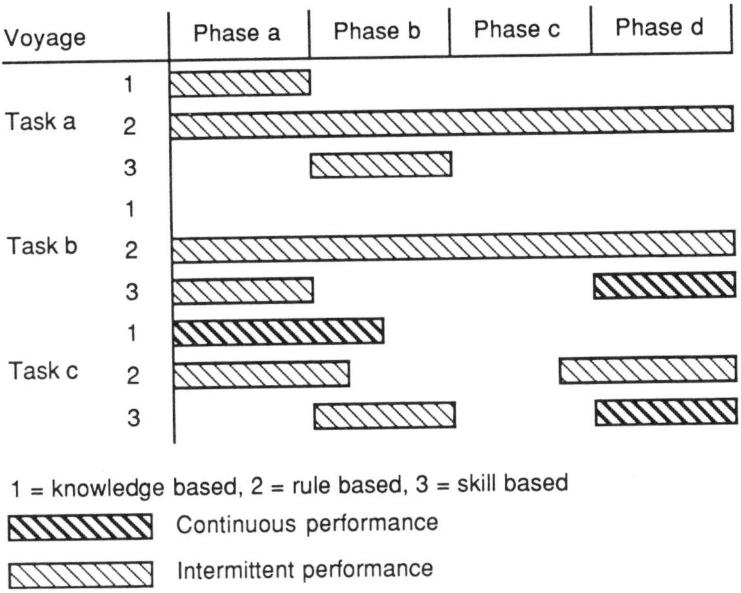

Voyage | Phase a | Phase b | Phase c | Phase d

Task a 1 2 3

Task b 1 2 3

Task c 1 2 3

1 = knowledge based, 2 = rule based, 3 = skill based

Continuous performance

Intermittent performance

Figure 17.4. The normative task model

The Generic Error Modelling System, as developed by Reason, was applied to the normative bridge task as a predictor of human error in order to use the results for the formulation of design requirements (Hale and Stoop, 1988). Verification of the results was achieved by a comparison with accidents registered by the Dutch Shipping Council and by on-the-spot observation on the bridge of a fishing vessel. This showed that the GEMS model explained 26 (72%) of the 36 collisions (Heinrich, 1988b).

Solutions

Solutions were proposed for problems of mental overload and underload but a number of other solutions were still needed. Solutions relating to training and regulation required a compulsory course in knowledge of regulations and use of equipment, a qualified replacement for the skipper, additional crew on the bridge in difficult circumstances, and installation of a watchman alarm (Draijer *et al.*, 1988).

General safety requirements for the design of the bridge

In these pilot studies, the formulation of safety requirements for the design of the bridge was restricted to the navigation and fishing tasks and to the scenario with normal use conditions. Both accident analysis and application of the human error theory resulted in a number of general design requirements (Heinrich, 1988c).

1 The average frequency of use of the instruments should be a determining factor to their positioning.

2 Grouping of instruments is useful to enable the comparison of information. Possible combinations are, for example,
   - decca and chart to transfer coordinates
   - radar and autopilot to detect and correct bias errors
   - compass, decca, radar and autopilot to compare information.

3 Modifications on the instrument level are possible:

   a automation of transfer of coordinates from the decca to the chart
   b integration of instruments; (with respect to the navigation, the radar and video trackplotter could be integrated; Obstacles could be stored in the memory of the plotter and would not be overlooked on radar)
   c linkage between instruments; (the decca and chart could be linked and the decca and autopilot could be linked for automatic changes in course. This is currently prohibited by the Shipping Inspectorate because of the hazards involved in automatically generated changes in course at preset course alteration points without regard to the traffic situation)
   d installation of alarms on instruments; (for the fishing process a 'guard' zone on the radar is desirable to warn against approaching ships during fishing gear operation). For the navigation equipment several alarms could be installed
      - a watchman alarm has recently been made compulsory
      - an alarm on the decca
      - an alarm on the combination of decca and autopilot to warn of arrival at course alteration points, incorporating an active go-ahead order from the watchman (this alarm is permitted by the Shipping Inspectorate)
      - an alarm on the trackplotter for known obstacles and programmed course alteration points
      - the introduction of ARPA; (at present this equipment is far too expensive, moreover, ARPA requires an independent feedback and is sensitive to bias errors; before introduction it needs to be absolutely reliable and requires the development of trained staff)
      - a radar alarm to detect an approaching radar signal
      - a collision alarm
   e introduction of a hierarchy of alarms, related to existing alarms for other functions such as engine, power supply, communication, fish hold and ship conditions
   f replacement of obsolete equipment such as decca by a video trackplotter.

4 Vision could be improved by illumination of decks and equipment on the bridge. Sight lines for several tasks with respect to navigation, decks and fishing gear could be improved.

5 Communication by speech and gesture to and from the crew should be improved.

6 Detailing of equipment could be improved. Principles of informational ergonomics need to be applied to instruments to improve readability and comprehensibility. The recognition of and distinction between controls for different equipment and functions should also be considered.

A new solution: the low-cost collision alarm

Due to rapid developments in micro-electronics and information technology, it is possible to reduce the chance of collisions by the installation of a device, linked to the radar, which sets off an alarm if another ship is coming too close. Such an instrument (ARPA) is already compulsory on ships of more than 10,000 tons, but is very expensive. The fishing industry is an outstanding target group for extension of the use of this sort of instrument, but it needs to be much cheaper. It has proved possible to develop such a device, although the dimensioning of the filter parameters is of the utmost importance for the prevention of false tracks by clutter at a short distance from the user ship (van der Sloot, 1988). The most decisive factor in its success is the reliable functioning of the radar system and the collision alarm.

Residual risks

If such an instrument were to be introduced, a number of residual risks and side effects must be reckoned with (van der Sloot, 1988). The residual risk is determined to a great extent by the reliability of the collision alarm. The solution demands a perfect functioning of radar and collision detector for credibility reasons, and it is an open question whether man or machine should be in control, a situation comparable with that in aviation. The device serves no purpose if the bridge is unmanned. Among younger generations of mates, 'blind' navigation by radar is more the rule than the exception. It should be considered whether it is possible to compel the mates to make a periodic check of the information presented by the equipment.

Finally a new accident scenario is introduced. It would be possible to install a navigation alarm on the deck, triggered by the collision alarm and the bridge might then remain unmanned until a warning was given. If the steering manoeuvres were also taken over by an autopilot, dangerous situations could occur during intense deck activities or during a period of rest after a heavy trawl. Blind faith in the instrument, supplemented by fatigue, could cause a low level of attention. An emergency situation might then cause panic and incorrect intervention if it was not clear which event had led to an alarm.

Final remarks

After a period of growth, the fishery industry in The Netherlands faces a period of cost reduction and quality improvement. Hull optimization for fuel reduction and sea performance, workplace design on the bridge and deck, processing of the catch and containerisation of the storage are all under discussion. A favourable climate is growing for improvements in bridge layout and equipment, although every Dutch skipper considers his bridge layout to be the best, and there is still much work to be done. This includes: information analysis of subtasks, definition of other tasks and of hazardous scenarios, integration of safety requirements with other design requirements, design and testing of mockups in cooperation with users, initiation of cooperation with shipyards, training institutes and government (research) institutions.

References

Bos, B. (1988). *Automatische Verkeerssystemen*. Technische Universiteit Delft, afstudeerverslag Faculteit Elektrotechniek.

Draijer, J.M., *et al.* (1988). *Minder aanvaringen met vissersschepen*. Technische Universiteit Delft, Faculteit Technische Wiskunde en Informatica ii–11 projektgroep.

Gelderblom, G.J. (1988). *Scenarios in de ontwerpmethodologie*. Technische Universiteit Delft, vakgroep Veiligheidskunde.

Hale, A.R. and Stoop, J. (1988). What happens as a rule? Communication between designers and road users. *Proceedings of Traffic Safety Theory & Research Methods Conference*, Amsterdam, 26–28 April.

Hale, A.R., Quist, B.W. and Stoop, J. (1988). Errors in routine driving tasks: a model and proposed analysis technique. *Ergonomics*, **31**, **4**, 631–641.

Heinrich, J.P. (1988)a. *Menselijk falen in de boomkorvisserij: een onderzoek naar de oorzaken en de mogelijkheden tot preventie*. Netherlands Institute for Fishery Investigation (RIVO), TO **88–09**.

Heinrich, J.P. (1988)b. *De mogelijkheden tot het voorspellen van Menselijk Falen in de boomkorvisserij met behulp van 'Human Error'-theorieën*. Netherlands Institute for Fishery Investigation (RIVO), TO **88–10**, Doctoraalscriptie Rijksuniversiteit Leiden, vakgroep Psychologische Functieleer.

Heinrich, J.P. (1988)c. *Ergonomisch brugontwerp: de methodologie om te komen tot het Pakket van Eisen*. Netherlands Institute for Fishery Investigation (RIVO), TO **88–11**.

Janssen, W.H. (1986). *Modellen van de verkeerstaak: de 'state of the art' in 1986*. Rapport IZF-TNO, **C-7**.

Kjellèn, U. (1984). The deviation concept in occupational accident control. *Accident Analysis and Prevention*, **16**, **(4)** 289–306.

Quist, B.W. (1987). The application of human error models in accident hypothesis formulation. *Second International Conference on Road Safety, Groningen*, 31 August–4 September.

Raad voor de Scheepvaart, nr 17, (1984). Uitspraak inzake het ongeval aan boord van een vissersvaartuig, waarbij drie opvarenden werden gewond. *Bijvoegsel Nederlandse Staatscourant* 2 mei.

Raad voor de Scheepvaart, nr 15, (1988). Uitspraak inzake het kapseizen en zinken van een vissersvaartuig tijdens het vissen op de Noordzee. *Bijvoegsel Nederlandse Staatscourant* 15 juni.

Rasmussen, J. (1980). What can be learned from human error reports. In *Changes in working life*, edited by K. Duncan, M. Gruneberg, and D. Wallis. (London: John Wiley).

Reason, J. (1987). Generic Error-Modelling System (GEMS): a cognitive framework for locating common human error forms. In *New Technology and Human Error*, edited by J. Rasmussen, K. Duncan and J. Leplat. (John Wiley & Sons Ltd).

Regt, M.J.A.M. de (1987). Geluidniveaus aan boord van zeegaande kotters, dl 1, 2 en 3. *Technisch Physische Dienst TNO-TH*, (Delft).

Schuffel, H. (1988). *De scheepsbrug van de jaren negentig*. Studiedag van het Genootschap voor Veiligheidswetenschap, 's Gravenhage, 29 September.

Schuffel, H. and Boer, J.P.A. (1985). *De scheepsbrug in de jaren '90. Navigatieprestatie en mentale belasting van de wachtdoend officier*. Rapport IZF-TNO, **C-14**.

Sloot. B.J.R. van der (1987). *De noodzaak van een low-cost ARPA*. Technische Universiteit Delft, vakgroep Veiligheidskunde.

Sloot, B.J.R. van der (1988). *Verkenningen naar een low-cost ARPA*. Technische Universiteit Delft, afstudeerverslag Faculteit Elektrotechniek.

Stoop, J. (1987). The role of safety in the design process. In *New Methods in Applied Ergonomics*, edited by J. Wilson, E.N. Corlett and I. Manenica, (London: Taylor and Francis).

Veenstra, F.A. (1986). *Trillingsmetingen aan boord van de boomkorkotters GO 38 en GO 26*. Netherlands Institute for Fishery Investigation (RIVO), **TO** 86-04.

Wertheim, A.H. (1977). Explaining highway hypnosis: experimental evidence for the role of eye movements. *Accident Analysis and Prevention*, **10**, 111-129.

Wertheim, A.H. (1982). Droomrijden, waarom we achter het stuur in slaap vallen. *Psychologie*, mei, 18-23.

Chapter 18
Effects of changes in delivery of dairy products on physical strain of truck drivers

C-H. Nygård and J. Ilmarinen

Institute of Occupational Health, Department of Physiology, Laajaniityntie 1, SF-01620 Vantaa, Finland

Introduction

Dairy products constitute the largest food products group in Finland. Since about one-third of all foods are dairy products, some 200,000 kg are handled annually in the average store. The most relevant problem in delivery is the transport of goods, which is mostly performed manually. Truck drivers deliver from 4,000 to 19,000 litres of dairy products manually each day. Their work includes lifting, carrying and pulling baskets. New packing methods and delivery systems have been introduced to facilitate and increase the delivery of large volume liquid food products. A rolling delivery system for dairy products has been developed, allowing the products to be moved from the dairy to the truck and later, from the truck to the refrigeration room of the store, on transport dollies. The aim of this study was to determine how the physical strain of the work has been affected by the change in transport method.

Materials and methods

Ten truck drivers between 22 and 51 years of age (mean age 35 years) participated voluntarily in the study. Their job consisted of loading the truck at the dairy, driving the truck and delivering the products to the stores. They had an average of eight years of work experience with the former method and about six months with the new method.

The drivers' physical working capacity was measured in the laboratory. This included the direct assessment of maximal oxygen consumption on a bicycle ergometer and anthropometric measurements (Table 18.1). The subjects' VO_{2max} was retested both before and after the field measurements to control any changes in physical working capacity. There were no significant differences between the working capacity in the baseline and the follow up measurements.

The subjects were measured during their work on two days in one week (a so-called light and a heavy day) both before and after the change in transport method. During the whole work shift, heart rate was registered minute by minute with an ambulatory device

Table 18.1. Age, anthropometry, maximal heart rate (HR_{max}) and maximal oxygen consumption (VO_{2max}) of ten truck drivers

	Age (yrs)	Weight (kg)	Height (cm)	Fat %	HR_{max} (beats. min^{-1})	VO_{2max} (ml. min^{-1}. kg^{-1})
Mean	35.4	79.4	176.7	20.5	180.7	36.5
SD	10.9	9.5	6.3	3.3	10.4	5.1

(Howel Corder; Rutenfranz *et al.*, 1977). Parallel to the heart rate recordings, observation of the dominant work tasks, extreme working postures of the back and quantity of goods handled were registered minute by minute. The oxygen consumption at work was measured for the most typical work tasks during both loading and delivery (Morgan Oxylog; Louhevaara *et al.*, 1985). The field studies were carried out in central Finland (Jyväskylä) during the winter (from January to April) both before and after the change in transport method. At that time there was snow on the ground, and the outside temperature ranged between minus 20°C and plus 2°C (mean temperature minus 8°C).

Results

The new delivery method increased the time of pushing and pulling by 2.8% during the loading phase and by 2.3% during the unloading phase (Table 18.2). The time of carrying goods decreased by 0.8% during loading and 2.2% during unloading. Driving the truck was the longest task, and the duration was increased by the new delivery method. The amount of dairy products carried, in litres per day, decreased by 55% with the new method, whereas the amount of pushing increased by 7% (Table 18.3)

Table 18.2. Mean time (of ten drivers, in percentage of work shift) used for various tasks in the former and the new transport method

Work task	Former method	New method
Loading		
Pushing and pulling	3.0	5.8
Carrying	1.6	0.8
Unloading		
Pushing and pulling	6.3	8.6
Carrying	19.8	17.6
Handling of empty baskets	5.2	4.0
Driving truck	35.2	41.1
Walking without burden	12.5	15.9
Pause	15.5	5.3

Table 18.3. Dairy products handled (gross litres/per day/per person) by work task and transport method

Work task	Former method	New method
Carrying	1,266	575
Pushing	4,685	4,998
Total	**5,951**	**5,573**

Oxygen consumption

During the former method the highest oxygen consumption occurred when pushing and pulling while loading. During the new method, carrying while loading required the highest oxygen consumption (Table 18.4). Oxygen consumption during the carrying was higher with the new method than during loading with the former method ($p<0.05$). During the unloading of goods, both pushing and carrying were decreased by the new method, although the differences were not statistically significant.

Table 18.4. The oxygen consumption in different work tasks of ten truck drivers

Work task	Oxygen consumption ($ml.min^{-1}.kg^{-1}$)			
	Former method		New method	
	Mean	SD	Mean	SD
Loading				
Pushing	21.2	4.5	19.7	2.8
Carrying	17.2*	2.6	21.1*	1.1
Unloading				
Pushing	17.0	0.6	14.9	1.5
Carrying	19.9	3.9	18.2	2.4

* Difference between the former and the new method statistically significant, $p<0.05$.

Cardiorespiratory strain

The highest mean heart rate during the former method occurred when loading the truck by pushing (120 beats.min^{-1}), during the new method when loading the truck by carrying (119 beats.min^{-1}) (Figure 18.1). There were no statistically significant differences in mean heart rates between the former and the new method.

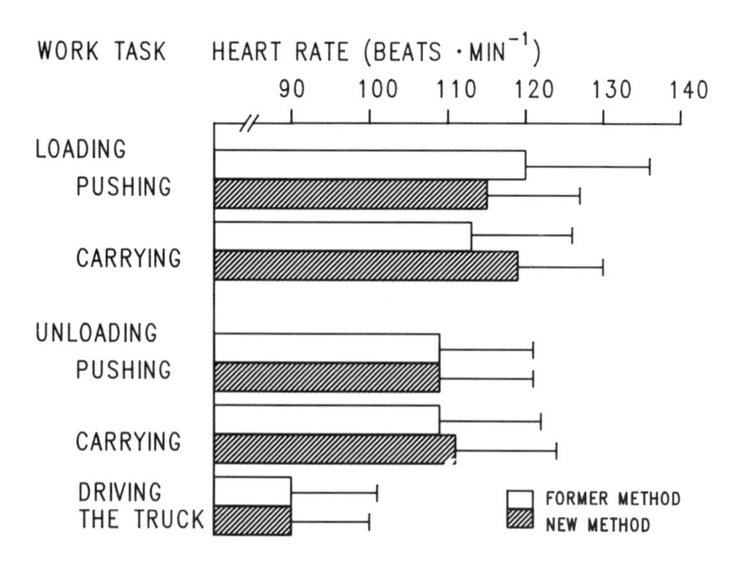

Figure 18.1 Mean and standard deviation of the heart rate of ten drivers during different work tasks for the two working methods.

The cumulative distribution of time in different heart rate classes shifted slightly to the right with the new method, indicating higher cardiorespiratory strain, whereas there were more peak loads above 120 beats.min^{-1} when using the former method (Figure 18.2). The differences were, however, not statistically significant. The relative aerobic strain (RAS) decreased by about 5% during loading with the new method but increased by about 8% when carrying (Figure 18.3). The new method also decreased the RAS during carrying when unloading by about 5%. The differences were, however, not statistically significant.

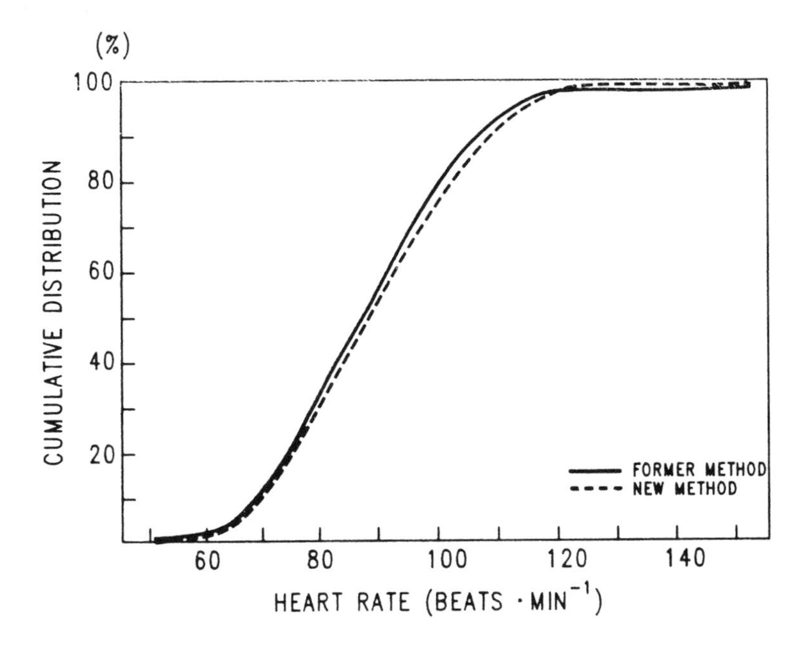

Figure 18.2. The cumulative distribution of time, (% of work shift) at different heart rates for the two working methods (mean of ten drivers).

Poor working postures

The incidence of extreme postures of the back was about 45% less during loading and about 39% less during unloading with the new transport method as compared to the former method (Table 18.5).

Discussion

The introduction of carts into jobs involving mainly lifting and carrying decreased workers' physiological strain (Hansson and Klusell, 1977; Oja *et al.*, 1979). A container or dolly should be even more favorable than a cart (Martin *et al.*, 1980). Our results indicated, however, that the new rolling method decreased the physiological strain only during loading at the dairy and did not affect the strain during unloading the truck at the markets

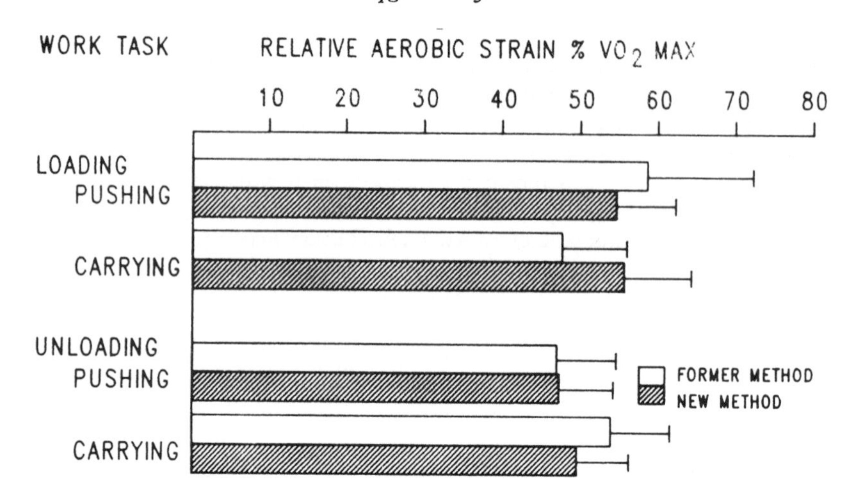

Figure 18.3. The relative aerobic strain in % VO_{2max} (mean and standard deviation) during different work tasks for the two methods.

Table 18.5. Number of minutes spent in extreme working postures of the back by work task and transport method

	Former method	New method
Loading	42	23
Unloading	205	126
Total	**247**	**149**

-deep forward bending (>90)
-back strongly twisted
-continuous backlift (>30 sec)

and stores. This finding is a result of the better facilities in the new dairies which were built for the use of carts and dollies; no improvements were made for the goods delivery at the stores. The stores were often situated in old buildings, and many of the smaller ones had no platform for unloading goods. Also, the route from the truck to the store often had stairs and thresholds. Although centralisation of small food shops to bigger markets has been a recent trend, there are still many small shops with very poor access for carts and dollies.

With the new method the workers handle more goods at a time in the dolly, than while carrying. The handling of a heavy dolly requires more static muscular effort, especially if the ground is slippery or if there are obstacles in the way (e.g. snow, thresholds). Static muscle work increases heart rate more than oxygen consumption, and this may explain why the mean heart rate during work had not changed. Although the new rolling method has increased the use of dollies and carts, the truck drivers still spend about 18% of the work shift carrying goods, and only 14% of the shift using carts and dollies.

The relative aerobic strain was, on average, 49% of VO_{2max} during loading and unloading, which is about the same as for dustmen pushing waste containers (Klimmer *et al.*, 1982) or men carrying mail (Oja *et al.*, 1977). A limit of 50% VO_{2max} has been suggested for heavy dynamic work with breaks (Andersen *et al.*, 1978) and the strain of delivering dairy products is close to that level. The most strenuous manual material handling tasks were during loading and unloading, which together account for about 35% of the work time. These tasks occur at intervals between driving the truck, which is light work according to heart rate (89 beats.min^{-1}) and oxygen consumption (0.4 l.min^{-1}).

In addition to high cardiovascular strain, the muscular strain is high. Hedberg (1987) found a significant increase in the serum creatinine kinase level among truck drivers at the end of the week, indicating high muscular strain. The high strain among the drivers was also combined with an elevated risk of musculoskeletal complaints.

The maximal oxygen consumption of the truck drivers (mean age 35 years) was about the same (36.5 ml.kg^{-1}.min^{-1}) as that of transport workers (mean age 52 years) in the municipal sector in Finland (Suurnäkki *et al.*, 1985). This means that the physical working capacity is rather low when compared with the work demands. The truck drivers' heavy work has not maintained or improved their physical working capacity. The same results have been found also in different types of industrial work (Ilmarinen *et al.*, 1981) and in auxiliary work (Ilmarinen *et al.*, 1985).

The results indicated that the physiological strain during the delivery of dairy products has been changed only a little by the new rolling delivery method. The beneficial effects of pushing the dollies will become visible only after marked improvements have been made for receiving the dollies in the stores. This study emphasized that technical changes in manual goods transport can affect the strain, but that the achievement of a marked decrease in strain requires ergonomic improvements in the entire delivery chain.

References

Andersen, K., Rutenfranz, J., Masironi, R. and Seliger, U. (1978). Habitual physical activity and health. *WHO Regional Publications: European series*, **6**. World Health Organization.

Hansson, J.-E. and Klusell, L. (1977). *Ergonomiska specialstudier av kärrning och bärning av sopor*. (Arbetarskyddsstyrelsen, Undersökningsrapport 1977, 45 s).

Hedberg, G. (1987). Epidemiological and ergonomic studies of professional drivers. *Arbete och Hälsa*, **9**.

Ilmarinen, J., Rutenfranz J., Kylian, H., Klimmer, F., Ahrens, M. and Ilmarinen, R. (1981). Untersuchungen über unterschiedliche präventive Effekte von habituellen körperlichen Aktivitäten in Beruf bzw. In der Freizeit auf die kardiopulmonale Leistungsfähigkeit. *International Archives of Occupational and Environmental Health*, **49**, 1–12.

Ilmarinen, J., Luopajärvi, T., Nygård, C-H., Suvanto, S., Huuhtanen, P., Järvinen, M., Cedercreutz, G. and Korhonen, O. (1985). Kunnallisten työntekijöiden toimintakyky (Functional capacity of municipal employees). *Työterveyslaitoksen tutkimuksia*, **3**, 212–238.

Klimmer, F., Kylian, H., Rutenfranz, J. and Grund-Eckardt, R. (1982). Belastungs–und Beanspruchungs–Analyse verschiedener tätigkeiten bei der Müllabfuhr. *Zeitschrift für Arbeitswissenschaft*, **36**, 90–94.

Martin, G., Frauendorf, H., Erdmann, U., Köhn-Seyer, G. and Vildosola, J. (1980). Arbeitsphysiologische Untersuchungen an Mülladen während der Hausmüllberäumung unter Verwendung von 110–Liter–Mülltonnen und Müllgrossbehältern. *Zeitung ges Hyg.*, **26**, 579–582.

Louhevaara, V., Ilmarinen, J. and Oja, P. (1985). Comparison of three field methods for measuring oxygen consumption. *Ergonomics*, **28**, 463–470.

Oja, P., Louhevaara, V. and Korhonen, O. (1977). Age and sex as determinants of the relative aerobic strain of nonmotorized mail delivery. *Scandinavian Journal of Work, Environment and Health*, **3**, 225–233.

Oja, P., Ilmarinen, J. and Louhevaara, V. (1979). *Postin kantomenetelmien fyysinen kuormittavuus* (Physical strain in mail delivering). (Työterveyslaitos, Helsinki, 32 s.).

Rutenfranz, J., Seliger, V., Andersen, K., Ilmarinen, J., Flöring, R., Rutenfranz, M. and Klimmer, F. (1977). Erfahrungen mit einem transportablen Gerät zur kontinuerlichen Registrierung der Herzfrequenz für Zeiten bis zu 24 Stunden. *European Journal of Applied Physiology*, **36**, 171–185.

Suurnäkki, T., Nygård, C-H., Ilmarinen, J., Peltomaa, T., Järvenpää, I., Järvinen, M., Nieminen, K. and Huuhtanen, P. (1985). Työntekijöiden kuormittuminen kunnallisissa ammateissa (Stress and strain in municipal work). *Työterveyslaitoksen tutkimuksia*, **3**, 239–261.

Chapter 19
Application of ergonomic principles to press shop operations in the light engineering industry

Harsh Vardhan Bhasin, K.P. Kothiyal and Amitabha De
National Institute for Training in Industrial Engineering, Vihar Lake, Bombay 400 087, India

Introduction

Press shops form an integral part of light engineering industry, where a variety of components are produced in semi-automatic or manual operations. In terms of workloads, most of the press shop operations are considered to be light in nature. The components handled by workers weigh only tens or hundreds of grams. However, the workers frequently interrupt their work and take unscheduled rests. This tendency is often attributed to lack of motivation and/or monotony of the work. While lack of motivation may at times exist in some workers, serious investigation is needed to identify the real causes.

This study was undertaken with the aim of identifying the design features of the work/workstations that may hinder the various functions carried out by the machine operators. An attempt was made to propose alternative workstation designs or eliminate the problems, and some proposed workstation modifications were tested in the laboratory.

Study of press operation

Location

The study was carried out in the actual working environment of a local light engineering factory. The factory managers had reported that the workers in the press shops were taking unscheduled breaks, thereby affecting the output targets.

Job classification

All the jobs done by the workers in the press shop were classified into five categories (as shown in Table 19.1). The classification was based on the weight handled and the postures adopted by the workers. Thus, three classes were obtained in the sitting and two in the standing posture.

The jobs consisted of picking up materials from the storage locations and then feeding them into a die. After proper positioning of the material, a switch or lever was pushed

Table 19.1. Classification of jobs according to working posture and weight handled by the operator

Working posture	Seated operator	Standing operator
Forward feeding	Class I 50g (Forcep fed)	Class V 5–10kg (Hand fed)
	Class II 100–500g (Hand fed)	
Sideward feeding	Class III 100–500g (Hand fed)	Class IV 100-500g (Hand fed)

to actuate the press. The 'C' type presses were pneumatically controlled. The workcycle times for the jobs varied from 6 seconds to 24 seconds. The materials were either metal strips, plates, or small components which were fed from the front or from the side. The material locations in the case of job classes I and II were in front of the operator, at table level. In other cases, materials were placed either at the sides and at the table level (classes III, and IV) or within a tote box at ground level (class V).

Subjects

Three operators, aged 29, 30 and 33 years, were selected for the experiments. They were physically fit and had no serious orthopaedic or other medical problems. They had been working in the factory for several years and had at least eight to nine years of work experience in the press shop. The purpose of the study was explained to them and their consent was obtained. The subjects were requested to work as usual on the machines, and each worked at all the jobs.

Measurements

Heart rate was taken at regular intervals (30 minutes) using a stethoscope. Work and rest patterns were recorded for the jobs. Time lapse photography and discomfort rating charts were used to identify the zones and extent of local discomfort.

Results and discussion

Table 19.2 shows the heart rate increments, discomfort zones and relative severity of discomfort for different classes of jobs. The δHR values shown in Table 19.2 are calculated as the difference between the working heart rate obtained during the work and the resting heart rate. The average resting heart rates for the subjects in the sitting and standing postures were 72 beats/minute and 80 beats/minute respectively. The jobs belonging to classes I, II and III did not show very much change in the heart rate above the resting level. However, in contrast, the jobs in classes IV and V led to large increases of 16–34 beats/minute.

In no case did the heart rate exceed 110 beats/minute, so that, in terms of workload, all the jobs could be classified as light (Christensen, 1953). However, the workers felt that the jobs were heavy and led to pain. As indicated in Table 19.2, the discomfort ratings

Table 19.2. Working heart rate increment, discomfort zones and severity of local discomfort for different classes of job

Component and job classes	Press size	δHR (beats/minute)	Discomfort zones (and levels*) after 4 hours of work
Individual light component Class I	25 ton	0–2	Upper back, thumb and forefinger (2)
Individual medium component Class II	40 ton	2–6	Upper back, thumb and forefinger (2)
Strips Class III	65 ton	4–8	Upper arm, shoulder,upper back (3), forearm, thigh, thumb and forefinger(2)
Strips Class IV	100 ton	16–20	
Heavy Plates Class V	100 ton	30–34	Right shoulder, arms, leg (2) Back–upper, mid (2), lower (3)

* 1 = Moderate, 2 = Moderately Heavy, 3 = Heavy, 4 = Unbearable
δHR is increment of work heart rate over resting heart rate.

varied from moderately heavy (score 2) to heavy (score 3). In the side-feeding tasks (classes III and IV), in both standing and sitting postures, the zones of heavy discomfort were the upper arm, shoulders and upper back. Discomfort was also experienced in the forearm, thigh, and hand, but was rated as moderately heavy. In the standing, front-feed situation (class V), the lower back was the zone of most discomfort.

In order to investigate why workers experienced pain and discomfort, despite the work being light, the postures most frequently adopted by the workers were examined. These were recorded at regular intervals using time-lapse photography, and revealed that workers were being forced to adopt uncomfortable postures due to the design features of the machines. In the side-feed situation, the worker faced the machine directly and thus had to stretch his left arm to pick up and feed the material. As Chaffin (1987) has pointed out, this is the least efficient position for force application. Moreover, as the 'C' type presses in use in the factory did not provide enough leg room, the operators were forced to sit at a distance from the press, which increased the forward reach. As a consequence, workers had to adopt a forward bent posture while positioning the material on the die, as illustrated in Figure 19.1. The job was therefore perceived as heavy due to the compounding effect of forward bending and arm stretching.

In the case of the standing, front-feeding job, a heavy postural load was imposed on the workers due to inappropriate material location. In this job, the material (heavy plates) was brought to the workers on a pallet truck in a tote-box which was placed low down. The workers had to bend down to ground level to pick up the plates and then front-feed them into the press (again shown in Figure 19.1). After the press operation, the plates were picked up and placed onto another tote-box. An additional problem was encountered in this job. Since a heavier press (100 tons) was used and the size of the platform on which the plates were placed was also proportionally larger, the resulting distance between the operator and the centre of the die was found to be large, necessitating a forward bent posture in order to position the plates properly on the machine. The postural load was thus considerable and as a result the workers experienced discomfort.

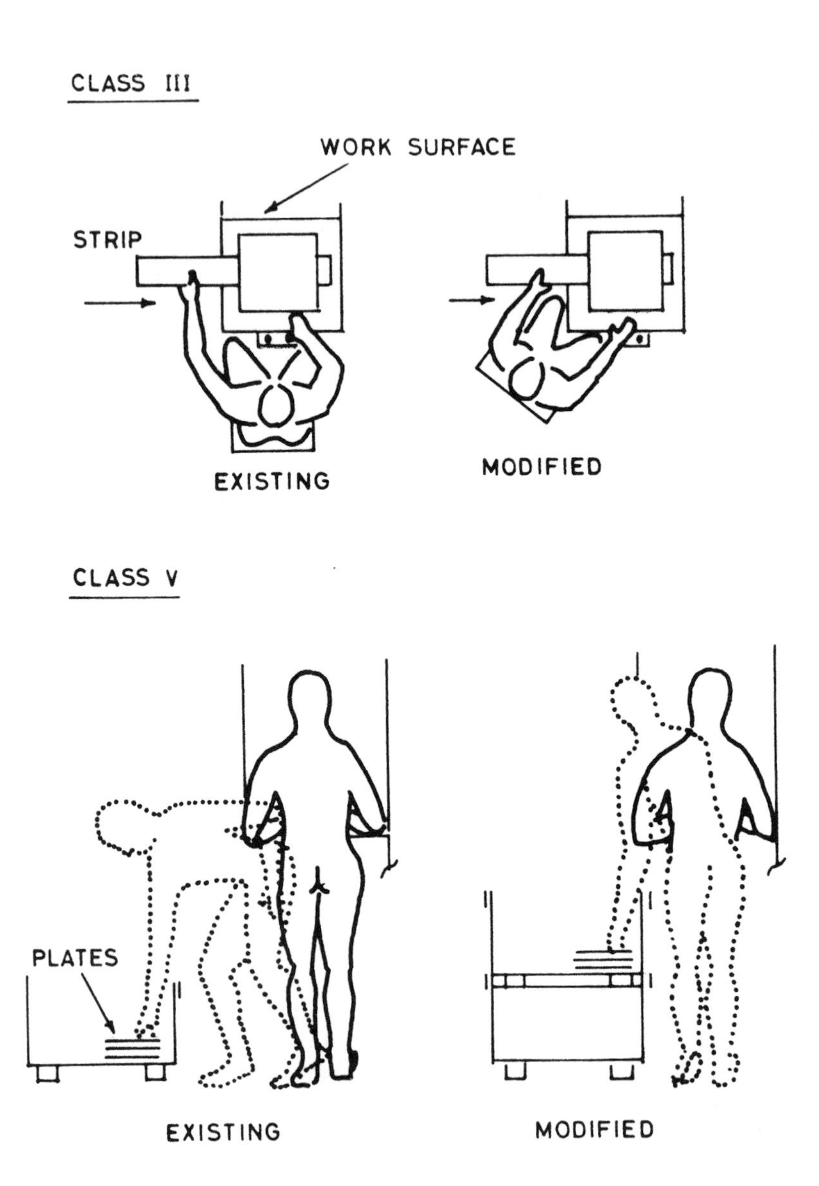

Figure 19.1. Working postures in existing and modified workstations

Experimental trials

On the basis of the discomfort ratings, two jobs were selected for modification. These were class III (side feeding, sitting) and class V (front feeding, standing). These jobs were simulated in the laboratory, and the results of the experiments are shown in Table 19.3.

In the sitting job (class III), in which strips were fed from the side, the operator was asked to sit obliquely (at an angle to the machine) so that side-bent postures could be avoided to some extent. Forward strip feeding would have been the most desirable solution but that would have necessitated redesigning the die. As this was not possible, it was felt that improvement could be achieved by reorienting the operator.

In the case of the standing, front-feed job, the tote-box height was raised from the floor level to about 70 cm so that the operator could avoid full bending while lifting the plates.

Results

Table 19.3 clearly indicates that the modifications made to the workstation design led to improvements in the workers' performance. The modifications improved cycle time, reduced work interruptions and lowered the heart rates.

Table 19.3. Work interruption patterns and incremental heart rates for existing and modified workstations

Component and job classes	Duration of work (minutes)	Number of parts produced	δHR (beats/minute)
Strip (existing) Class III	11.8	22	6
Strip (modified) Class III	22.7	49	4
Plate (existing) Class V	10.0	15	30
Plate (modified) Class V	13.7	30	22

Note: rest duration was normally 3 to 4 minutes

Discussion

In the actual working conditions, the discomfort/pain experienced by the workers due to the postural loads seemed to be indicated by frequent interruptions of work, with breaks of at least 3–4 minutes. The rest periods were recorded during the work, but no consistent pattern could be observed. Although the frequency of work interruption depended on the type of job, it was not certain whether the interruptions occurred due to discomfort or to other reasons such as boredom or lack of motivation.

The problem was therefore further investigated. The results of the study simulated in the laboratory showed how long a well-motivated worker would take before he interrupted his work. The subjects were asked to perform the job continuously without breaks until they felt that they could not continue further. They were then allowed a break of 3–4 minutes before restarting the work. The resultant durations of work are given in Table 19.3.

This shows that minor ergonomic modifications made in work postures could result in considerable improvements in productivity and duration of continuous work. A reduction was also noticed in heart rate increment, especially for the class V job. Responses to the personal interviews with the operators revealed several reasons for rest pauses during work. These were as varied as an urge to smoke, a visit to the toilet or a feeling of physical discomfort. All the workers, however, agreed that the modified work stations were more comfortable.

It should be noted that this study was restricted to modifications in working postures, possibly with minor adjustments to the work place. There is considerable scope for improvement in workers' productivity if further modifications could include ergonomic aspects of the press and die design.

References

Chaffin, D.B. (1987). Biomechanical aspects of workplace design. In *Handbook of Human Factors*, edited by G. Salvendy (New York: John Wiley and Sons, Inc.), 601–619.

Christensen, E.H. (1953) Physiological valuation of work in Nykroppa Iron Works. In *Symposium on Fatigue*, edited by W.F. Floyd and A.T. Welford (London: H.K. Lewis).

Chapter 20
Ergonomic guidelines for adjustment and redesign of sewing machine workplaces

Nico J. Delleman and Jan Dul

TNO Institute of Preventive Health Care, Musculoskeletal Research Group, P.O. Box 124, 2300 AC Leiden, The Netherlands

Introduction

Numerous studies have shown that operators of sewing machines report discomfort in the left shoulder, the neck and the back (e.g. Vihma *et al.*, 1982). These complaints may be caused or aggravated by the working posture which is characterised by an elevated left upper arm position and a flexed position of the head and trunk. At a traditional sewing machine workplace, the body posture is constrained by (1) the eyes for visual control of the work, (2) the hands to direct the sewing material, and (3) the foot to control the speed of the machine. In order to improve the working posture and reduce the number of complaints, quantitative ergonomics guidelines for adjustment and redesign of sewing machine workplaces are needed which take these postural constraints into account. At present, only a preliminary guideline for table height at a sewing machine workplace exists. Dul *et al.* (1990) found that a table height 5 cm above elbow height gave a better working posture than lower tables. Higher tables are expected to improve the head and trunk posture, but are also likely to create a more elevated position of the left upper arm. To resolve this classic problem the table desk can be given a slope. This may result in a more upright posture of the head and trunk, without elevating the arms and loading the shoulder region any further, as seen in various settings (e.g. Bendix and Hagberg, 1984). Wick and Drury (1986) have shown the positive effect of an inclined table desk on the working posture for sewing at a so-called single-post sewing machine in the garment industry. No guidelines are available for the position of the pedal in the fore/aft direction.

The purpose of the present study was to formulate guidelines for adjustment and redesign of sewing machine workplaces in order to improve working posture and to reduce the load on the musculoskeletal system. The relevant variables are: table height, table slope and pedal position.

Methods

Experimental design

Ten experimental conditions consisting of different combinations of table height, table slope, and pedal position were tested (Figure 20.1). Industrial sewing machine operators worked for a certain period in each of them.

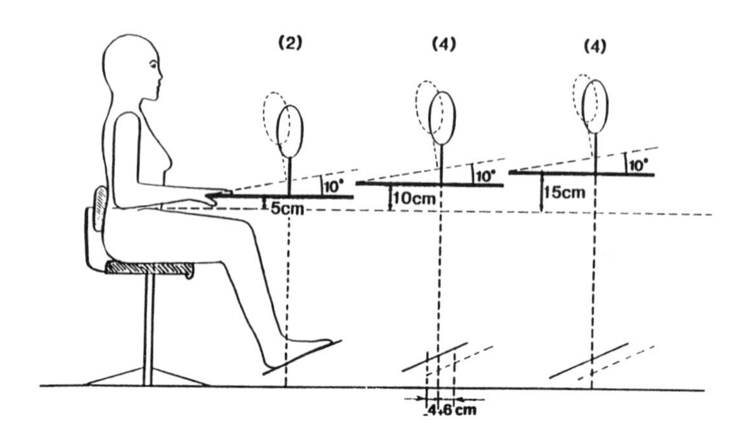

Figure 20.1. The ten experimental conditions. The elbow height is shown by the horizontal broken line. The needle position is shown by the vertical broken line.

Independent variables

1. Table height: +5, +10, and +15 cm relative to elbow height. Elbow height was defined as the distance from the floor to the elbow with the operator sitting upright, looking forward, the upper arms hanging down, and the forearms horizontal (Figure 20.1).
2. Table slope: table desk 0 and 10 degrees inclined to the horizontal.
3. Pedal position: −4 cm, i.e. the axis of the pedal 4 cm to the operator's side of the needle (lateral view of the workplace), which is in accordance with the average pedal position at the operator's own industrial workplace, and +6 cm, i.e. 6 cm to the opposite side of the needle, which is 10 cm further under the table seen from the operator, compared to their own workplace.

A complete block design (3 × 2 × 2) turned out to be impracticable. In preparing the experiments, two experimental conditions (with table height +5 cm and pedal position +6 cm, table slope 0 and 10 degrees) appeared to cause problems due to lack of leg space, and were excluded. The remaining ten experimental conditions were tested.

Measurements

Four independent parameters were measured in each of the ten experimental conditions. Two of them were related to the working posture and two were related to the operators' own responses to questionnaires. The working posture of the operator was recorded by the opto-electronic VICON-system with four synchronized video cameras. Retro-reflective markers were put on the skin overlying selected body joints and bones. Data acquisition was restricted to the time intervals during which the machine was running (when the posture was maximally constrained), at the end of a working period. Next, markers were identified semi-automatically. Based on the three-dimensional positions of markers, the head inclination and the elevation angle of the left upper arm were calculated.

Head inclination

Markers were placed near the lateral corner of the eye and near the lobe of the ear. The sagittal plane projection of the angle between the line through these two points and the vertical was calculated for the working posture and for a neutral position (vertical arm, head and trunk upright). Head inclination was defined as the difference between the two calculated angles. It can be considered as the sum of head, neck and trunk flexion.

Elevation of the left upper arm

Markers were placed on the left acromioclavicular joint and on the insertion position of the deltoid muscle at the left upper arm. The angle between the line through these two points and the vertical was calculated for the working posture and for the neutral position (vertical arm, head and trunk upright). Elevation of the left upper arm was defined as the difference between these angles.

In addition the operators' own responses were recorded by two types of questionnaire:

Postural discomfort—total body score

A modified version of the method by Corlett and Bishop (1976) was used. Operators rated their postural discomfort in 30 regions shown on a diagram of the rear view of a human body. A six-point scale (0 = no discomfort, 5 = very severe discomfort) was used. This questionnaire was filled out at the beginning (B) and at the end (E) of a working period. On both, the scores on all body regions were summed. The resulting total score at B was subtracted from the resulting total score at E.

Comparison with own workplace

At the end of the working period the operators judged the experimental workplace in comparison with their own industrial workplace. A five-point scale (1 = much better, 2 = a little better, 3 = equal, 4 = a little worse, 5 = much worse) was used.

Subjects

Five female operators (average age 41 years, range 34–47; average stature 170 cm, range 167–175) from the furniture industry participated in the experiments. They were familiar with the sewing task (average experience 12 years, range 2.5–18).

Procedure

In the laboratory tests the operators worked at a traditional adjustable sewing machine workplace. They performed their normal sewing task in ten experimental sessions of 45 minutes followed by breaks of 15 minutes. In each session one of the ten experimental conditions was presented. The first day consisted of three sessions and the following two days of either three or four sessions. The order of presentation of the experimental conditions was balanced over subjects, days and sessions. Prior to the first session each operator selected a seat height at which the pedal operation was comfortable. For this, visual cues from the table desk and the sewing machine were hidden by a blanket. Next, the elbow height was measured. The seat height and pedal angle were constant during all experiments, but the subject was free to choose the horizontal distance from the seat to the table.

Data analysis

The effects of table height, table slope, and pedal position on all four dependent parameters were tested by an analysis of variance. The selected level of significance was $p = .05$.

Results

Head inclination

Figure 20.2a shows the average group effect of table height and table slope on the head inclination. A higher table caused a more upright position of the head and trunk. A table slope of 10 degrees also led to a more upright head and trunk posture than a horizontal table. Both effects were significant. The pedal position had no effect on the head inclination.

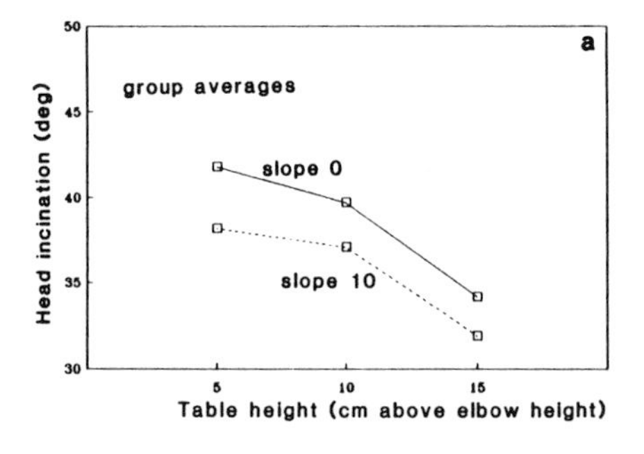

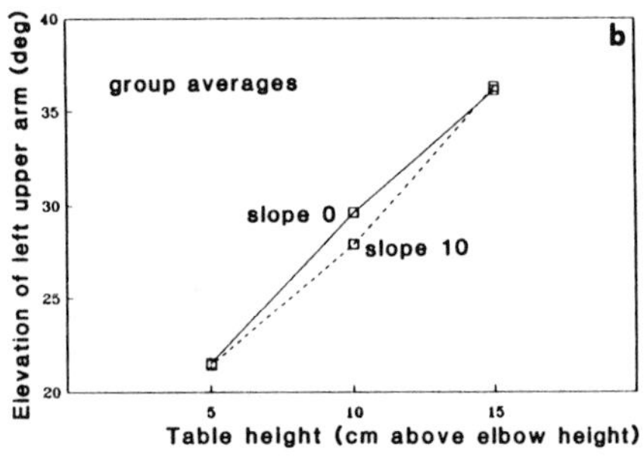

Figure 20.2. Average group scores for head inclination (a) and elevation of the left upper arm (b) in relation to the table height (relative to elbow height) and the slope of the table desk (relative to the horizontal).

Elevation of the left upper arm

Figure 20.2b shows the effect of table height and table slope on the elevation of the left upper arm. A higher table caused a more elevated position of the left upper arm. This effect was significant. Neither table slope nor pedal position had an effect on the elevation of the left upper arm.

Postural discomfort—total body score

Table 20.1 shows the average group scores for the ten experimental conditions. None of the independent variables showed a main effect on this dependent parameter. The combination of a 10 degree table slope and a pedal positioned further under the table (+6 cm) showed the best scores (scores −0.2 and −1.2 for table heights +15 and +10 respectively).

Comparison with own workplace

Table 20.1 shows the average group scores for the ten experimental conditions. None of the independent variables showed a main effect on this measure. The combination of a 10 degree table slope and a pedal positioned further under the table (+6 cm) showed a significant effect. On average the operators rated a workplace which was adjusted this way as a little better than their own industrial workplace (scores 2.0 and 1.8 for table heights +15 and +10 respectively).

Table 20.1. Average group scores for postural discomfort—total body score and comparison with own workplace of the ten experimental conditions

Experimental conditions										
Table height (cm)(1)	+15	+15	+15	+15	+10	+10	+10	+10	+5	+5
Table slope (deg)	10	10	0	0	10	10	0	0	10	0
Pedal position (cm)	+6	−4	+6	−4	+6	−4	+6	−4	−4	−4
Postural discomfort —total body score (2)	−0.2	0.6	1.6	2.4	−1.2	1.6	1.6	0.4	2.6	2.8
Comparison with own workplace (3)	2.0	3.3	3.8	3.8	1.8	3.5	3.3	3.0	4.3	4.0

(1) relative to elbow height, (2) amount of decrease or increase,
(3) 1 = much better, 2 = a little better, 3 = equal, 4 = a little worse, 5 = much worse

Discussion and formulation of guidelines

A higher table leads to a more upright and better posture of the head and trunk, and to a more elevated or worse position of the left upper arm. On the basis of these conflicting effects a single optimal table height can not be recommended. The two questionnaire parameters also do not show a group preference for a single table height. This is in accordance with the large variety of table heights (relative to elbow height) chosen by the operators at their own industrial workplaces (+5.5, +6, +9.5, +13, and +15.5 cm relative to elbow height). These individual preferences may be associated with different individual working methods. An operator who supports the arms on the table may prefer a higher table due to the positive effect on the head and trunk posture and the absence of the adverse effect of an elevated upper arm position on the shoulder load.

Dul *et al.* (1990) recommended that the table height should not be lower than 5 cm above elbow height. From the present study no recommendation for a single higher table height can be given. Therefore the first guideline for adjustment is that the table height has to be adjusted to at least 5 cm above elbow height. The specific height depends on the preference of the operator. Consequently the first guideline for redesign is that the table height has to be adjustable between the elbow height of a small percentile of the user population (in a comfortable working posture) plus 5 cm, and the elbow height of a large percentile plus at least 15 cm.

A table desk inclined 10 degrees towards the operator showed a positive effect on the head and trunk posture. This may be explained by the improved visual conditions. However, the magnitude of the postural improvement was not as large as that expected from change in the table slope. A reason for this may be that the operator would prefer a greater improvement, but the resulting more upright posture of the trunk would remove the elbow support on the table. To fulfill this need the trunk (and the head) would have to be bent forward again. The assumed need for elbow support is substantiated by the numerous table desk enlargements seen in industry. The positive effect of a 10 degree table slope on the head and trunk posture occurs irrespective of the pedal position. However, the two questionnaire parameters support this result only on the condition that the pedal is placed further under the table. Assuming that the operator feels the need to sit closer to the table (e.g. for elbow support without having to bend forward), only a combination of better viewing conditions, i.e. an inclined table desk, and the possibility of keeping the same comfortable posture of the lower extremities, i.e. a pedal positioned further under the table, will allow this. A pedal position further under the table requires enough free leg (knee) space, therefore, a somewhat higher table is to be recommended because of possible obstructions through various structures suspended below the table desk, e.g. the motor. The considerations above lead to a second guideline for adjustment: The table desk should be given a slope of at least 10 degrees and the pedal should be positioned as far under the table as considered comfortable. The table has to be raised if ample leg space is not available. Consequently the following guidelines for redesign emerge: The table desk has to be adjustable in slope, the pedal should be easy (free) to position, and ample leg space should be available.

References

Bendix, T. and Hagberg, M. (1984). Trunk posture and load on the trapezius muscle whilst sitting at sloping desks. *Ergonomics*, **27**, 873–882.

Corlett, E.N. and Bishop, R.B. (1976). A technique for assessing postural discomfort. *Ergonomics*, **19**, 175–182.

Dul, J., Baty, D., van der Grinten, M.P., Hildebrandt, V.H. and Buckle, P.W. (1990). The effect of table height on posture and discomfort of female sewing machine operators (in preparation).

Vihma, T., Nurminen, M. and Mutanen, P. (1982). Sewing-machine operators' work and musculo-skeletal complaints. *Ergonomics*, **25**, 295–298.

Wick, J. and Drury, C.G. (1986). Postural change due to adaptations of a sewing workstation. In: *The Ergonomics of Working Postures*, edited by E.N. Corlett, J. Wilson and I. Manenica, (London: Taylor and Francis), pp. 375–379.

Acknowledgements

The help of F.L. Piena during all phases of the study, and of M.P. van der Grinten in the construction of the questionnaires is gratefully acknowledged. The study was financially supported by the Directorate-General of Labour of the Dutch Ministry of Social Affairs and Employment.

Chapter 21
Precision manual assembly jobs: biomechanical considerations in workplace design

K.P. Kothiyal, H.V. Bhasin and Amitabha De

Ergonomics Research Laboratory, National Institute for Training in Industrial Engineering, Vihar Lake, Bombay 400 087, India

Introduction

In modern industrial set-ups workers usually perform light repetitive assembly-type jobs; these jobs mostly have to be done with skill and precision. A frequent practice in industry is to design a workplace for assembly jobs based on predetermined motion-time systems (PMTS). In PMTS workcycle times are determined by the number and the type (reach, grasp, move, etc.) of operations to be performed for the completion of a job. The workcycle times are optimized by eliminating all uneconomical, redundant and asymmetrical operations. For precision work, however, the PMTS method does not appear to be satisfactory. One of the limitations of PMTS is the lack of consideration of the posture at work (Chaffin, 1986). Posture plays a critical role in precision work (Laville, 1985), and the posture adopted by a worker on the job is largely determined by the design of the workplace. The present study was therefore undertaken to understand how work and workplace design affect workers' performance (workcycle time, endurance, etc.) and comfort. An electronic circuit assembly job, involving precision manual tasks, was selected for the purpose. The results are discussed from the biomechanical point of view. The paper also discusses the relative importance of various biomechanical parameters in workplace design.

Materials and methods

Two healthy male human subjects participated in the experiments. A work station for the electronic circuit assembly job was created in the laboratory. Printed circuit boards (PCBs), stored in front, were picked up by the operator with both hands and placed on a specially designed fixture on the table which held them in place during work. The fixture had the provision to hold the PCB at different inclinations with respect to the plane of the working table.

The circuit assembly job comprised inserting resistors into a PCB. Resistors were kept in a small tray and placed just in front of the operator. The position of the tray remained fixed in all the experiments. The job was completed in the following five steps: (1) picking up a resistor from the tray, (2) bringing the resistor to the pre-determined location on the

PCB, (3) aligning the resistor leads with respect to the holes on the PCB, (4) inserting the resistor into the holes by applying a little force and (5) bringing the hand back to the resistor tray to start the next cycle of work. The job was performed in the seated position. The working table height (71 cm) and the distance between the centre of the PCB and the backrest of the chair (51 cm) were kept constant during the experiments. These values were arrived at on the basis of the anthropometric characteristics of the subjects and their preferred working distances. Two patterns of resistor placement were considered. In Pattern I (PI), subjects were required to insert resistors into two parallel columns, each having 8 resistors. In Pattern II (PII), two parallel rows of 8 resistors each were formed. In addition to the two resistor patterns, the subjects worked at three PCB orientations. The angles of orientation were 0°, 15° and 30°. Thus each subject worked in six job situations and was asked to self-pace the work and continue as long as they could. No rest pauses were allowed during the work. Endurance time, workcycle time and the number of completed cycles were determined. The head position during work was also noted.

Results and discussion

Endurance and workcycle time

Table 21.1 shows the data on endurance time (T_{end}), mean workcycle time (T_w), and the number of completed workcycles in the two resistor configurations (Patterns I and II). The results clearly indicated that the pattern of resistor placement played a considerable role in determining workcycle time and hence the work output. Compared to PI, the mean workcycle time values in PII increased substantially; by about 25% for subject TMS and 13% for subject TR. Though the percentage changes for the two subjects differed somewhat, the trend was, however, consistent.

Table 21.1. Endurance time (T_{end}), number of cycles (N) and mean workcycle time (T_w) at PCB orientation angle = 0°. Numbers in brackets show standard deviations

	PI			PII			Percent change		
Subject	T_{end} (min)	N	T_w (min)	T_{end} (min)	N	T_w (min)	T_{end}	N	T_w
TMS	79.35	68	1.167 (±0.120)	70.18	48	1.462 (±0.121)	−11.6	−29.41	+25.2
TR	120.10	102	1.177 (±0.071)	122.57	92	1.332 (±0.095)	+2.0	−9.8	+13.1

PI = Pattern I, PII = Pattern II.

Effect of PCB orientation

Table 21.2 shows that mean workcycle time (T_w) for the resistor pattern PII is higher than PI for both the subjects in all the cases considered in the study. Table 21.2 also clearly brings out the effect of PCB inclination on T_w. As the angle of inclination was increased from 0° (horizontal position) to 15°, T_w decreased for both the resistor-patterns. Higher decrements (6–7%) were noticed for PII than for PI (about 3%). It was surprising to find that T_w registered an increase when a was increased from 15° to 30°. In both the subjects, the T_w value at $a = 30°$ was either nearly the same as in $a = 0°$ or somewhat lower, especially in one subject (TR) for PII. It appeared that two opposing factors affected the workcycle time. These factors were: (1) line of sight or vision and (2) steadiness of the forearm-hand segment. When the PCB was horizontal, the operator's line of sight was not in the same direction as the hole axis. This made the resistor insertion difficult. To be able to align the leads properly over the holes, the operator was forced to bend forward. The forward-bent posture adversely affected the workers' physical comfort and this was reflected in workcycle times. The forearm-hand segment steadiness in this case was found to be relatively good, but it still could not off-set the disadvantage due to the forward-bent posture. As the inclination angle was increased to 15°, the line of sight became closer to the hole axis. This, combined with the double point support to forearm-hand segment for most of the time during work, eased the insertion of the resistor leads into the holes. Thus the workcycle time decreased. However, when a was increased to 30°, the advantage of the double point support to forearm-hand segment was almost lost. In this case the elbow was lifted for most of the time and the subjects had to rely mainly on the support of the hand or fingers. In addition, obstruction due to the relatively high angle between the resistor leads and the hole-axis was encountered while pushing resistors into the holes. In order to overcome this, the operator had to perform an additional operation, i.e. rotating the resistor towards him/herself when both leads were over the holes. In this way a proper alignment between the hole-axis and leads was achieved for easy push-in. Consequently, higher workcycle times compared to $a = 15°$ were observed.

Table 21.2. Influence of PCB orientation (a) *on the mean workcycle time. Values in brackets are standard deviations*

Pattern	a	0°	15°	30°	Percentage change 0°–15°	Percentage change 0°–30°
PI	TMS	1.167 (±0.120)	1.133 (±0.060)	1.170 (±0.100)	−2.9	+0.25
	TR	1.177 (±0.071)	1.145 (±0.071)	1.166 (±0.099)	−2.7	−0.90
PII	TMS	1.462 (±0.121)	1.349 (±0.107)	1.451 (±0.123)	−7.7	−0.75
	TR	1.332 (±0.095)	1.248 (±0.065)	1.263 (±0.097)	−6.3	−5.2

Head movement during work

The importance of the head position in relation to the environment and the torso in high precision work has been emphasized by Laville (1985). In the present study therefore the role of head movement was investigated to find out how this affected a self-paced circuit assembly job. The head movement was monitored throughout the whole duration of work for all the job situations. Figure 21.1 presents the results of one subject (TR) for the first 30 workcycles. In the figure, head position changes are depicted for $a = 0°$, 15° and 30° while the subject formed the resistor Pattern I. Two interesting features of the head movement with regard to the mean head position and the range of the head movement during the job were observed. It was seen that the mean head position for all the three inclinations was different. The mean head position for the first 30 workcycles for $a = 0°$ was 10.34 ± 0.75 cm. At $a = 15°$ and 30°, the mean positions were 8.48 ± 1.22 cm and 8.70 ± 1.85 cm respectively. Thus the mean head position at $a = 0°$ was about 22% and 19% more forward bent than at $a = 15°$ and 30° respectively. This supports the contention above that the horizontal orientation of the PCB would necessitate a forward bent posture to insert the resistors into the PCB holes. Rohmert and Mainzer (1986) have also made similar observations with regard to posture in the horizontal arrangement of the PCBs in circuit assembly jobs. Laville *et al.* (1972) have investigated variations in the head position and the torso of the workers in machine-paced television assembly. Their observations also supported the results of the present study, as they found that the workers generally adopted a forward bent posture during work.

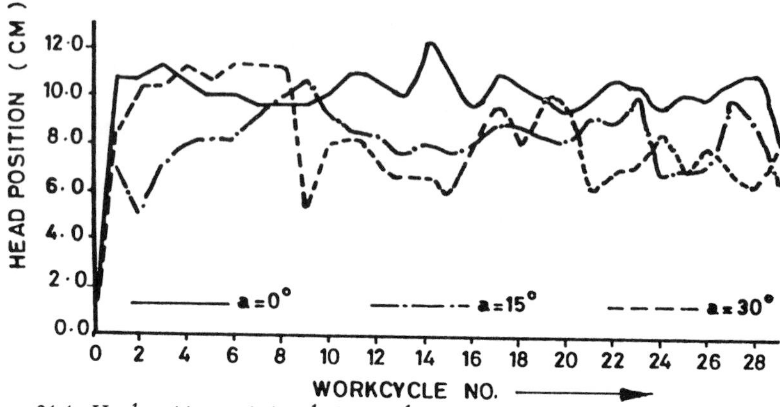

Figure 21.1. Head position variation during work

The second interesting feature, the range of head movement, clearly distinguished the three job situations. At $a = 0°$, the absolute range of movement was found to be relatively small (4.2 cm) compared to $a = 15°$(6.0 cm) and $a = 30°$(6.2 cm). The swinging of the head back and forth about a mean position is essential for minimizing discomfort in jobs of longer duration. Fixity of posture has been regarded by Grieco (1986) as an important factor contributing to discomfort and occupational health problems. Figure 21.1 shows that for $a = 0°$, the head movements were small and infrequent. In other words, the head remained fixed around the mean position for most of the time. At larger PCB angle orientations, on the other hand, frequent and larger head position changes were observed. Thus, job situations at higher angular orientations offered greater head mobility for postural adjustment during work.

Sites of body discomfort

An attempt was also made in this study to identify regions of body discomfort. The degree of discomfort in the body parts was assessed subjectively. The neck and back were singled out by both the subjects as the site of mild to heavy discomfort/pain in all the job situations considered in the study. One subject (TMS) reported severe pain in the shoulder region for PI at $a = 0°$. One important finding of the study was the significant reduction of discomfort at PCB orientations greater than 0°. A mild pain at the wrist joint was reported only at $a = 0°$. Subjects also reported mild eye strain, especially at the horizontal PCB orientation. In general, the subjects found working comfortable at $a = 15°$ and 30°.

Biomechanical considerations

From the biomechanical analysis viewpoint, the cervical joint (C6/C7), the mid-lumbar joint (L3/L4), and the shoulder joint appeared important as these were the main body regions where the subjects complained of pain. Considering the seated operator as a linked-segment system, the moments at those joints can be computed as:

$$M\ (\text{C6/C7}) = W(\text{H}) \times d(H)$$
$$M\ (\text{L3/L4}) = W(\text{L}) \times x(\text{L}) + W(\text{H}) \times x(\text{H}) + W(\text{UA}) \times x(\text{UA}) + W(\text{FA}) \times x(\text{FA})$$
$$M\ (\text{shoulder}) = W(\text{UA}) \times d(\text{UA}) + W(\text{FA}) \times d(\text{FA})$$

where M = moment; W's are the weights of the segments; H, L, UA and FA respectively denote head, upper lumbar, upper arm and forearm (including the hand); d's and x's are the perpendicular distances of the centre of gravity (C.G.) of the various segments from the joint about which moment is considered.

Equation 1 shows that the stress at the cervical joint is directly proportional to the angle made by the head with respect to the torso. Thus to reduce the pain in the neck, the PCB should be kept such that the head-torso angle is minimal. In the present study, this was achieved by orienting the PCB at angles 15° to 30°. Equation 2 indicates that the load at the lumber joint (L3/L4) is determined by the distance of the C.G. of the combined weight of the various segments. Forward bending would move the combined C.G. further away, thus increasing the moment at L3/L4. Therefore the forward bent positions at the workplace must be avoided. This can be done by bringing the PCB closer to the operator. At the shoulder joint, the moment is primarily determined by the forearm-hand position (Equation 3). Since the forearm-hand segment was mostly supported by the table during work, there was virtually no complaint of pain at the shoulder. Hence no change in the workplace design with respect to the shoulder joint was contemplated in the present study.

Conclusions

The main findings of the present study were:

1. Workcycle time (T_w) was found to depend on the angle of orientation a of the PCB; it was at a minimum for $a = 15°$.
2. Workcycle time was significantly affected by the pattern of resistor placement on the PCB. The column-wise pattern (PI) was found better than the row-wise pattern (PII).
3. Subjects normally adopted a forward bent posture, the degree of which varied with the PCB orientation, resistor pattern and the PCB distance from the backrest of the chair.
4. Subjects reported mild to heavy pain mainly in the neck and the middle lumbar regions. The degree of discomfort varied with PCB orientation and resistor patterns; maximum discomfort was experienced for PII at $a = 0°$.

References

Chaffin, D. (1986). Biomechanical aspects of workplace design. In *Handbook of Human Factors*, edited by G. Salvendy, (New York, Wiley Interscience), pp 601–619.

Grieco, A. (1986). Sitting posture: an old problem and a new one. *Ergonomics*, **29**, 345–362.

Laville, A. (1985). Postural stress in high speed precision work. *Ergonomics*, **28**, 229–236.

Laville, A., Teiger, C. and Duraffourg, J. (1972). Conséquences du travail répetitif sous cadence sur la santé de travailleurs et les accidents., *Report No. 29*, (Paris: Laboratoire de Physiologie du Travail-Ergonomie du CNAM).

Rohmert, W. and Mainzer, J. (1986). Influence parameters and assessment parameters for evaluating body postures. In *The Ergonomics of Working Postures*, edited by N. Corlett, J. Wilson and I. Manenica, (London: Taylor and Francis), pp. 183–217.

Acknowledgements

The authors are grateful to Dr S. Ramani, Director, NITIE for his constant encouragement. Thanks are due to Mr T.M. Sahu and Mr T. Ramaswamy for their help in conducting the study.

Part VI

Work organisation and job design

When trying to evaluate the effects of different ways of designing work we can run into many problems and have to make many compromises. Time scales can be not of our choosing; rarely can work design changes be evaluated before, during and after the change, and also some considerable time afterwards. Organisational imperatives or study resource restrictions usually curtail evaluation. The nature of our profession is such that often we are not given sufficient lead time in order to establish a true base level. The measures we should take comprise another problem area. Should we assess both subjective attitudes to change and also take so-called objective measures of shifts in production, quality and defects, downtime, direct and indirect costs, sickness, unexplained absence, labour turnover, accidents, etc? What mix of such measures should be applied? Most important, how easily can such measures be taken? Anyone who has studied companies' records on absenteeism or downtime, for instance, knows that these usually comprise partial and probably inaccurate data, never in the 'right' form for the needs of the work design analyst, and with a very short 'shelf life'.

Of course, the types of measurement used must reflect the type of change and the company's reasons for making it. The first of the papers in this section, by Šarić, Raboteg-Šarić and Takšić, compared two types of Yugoslavian assembly line: 'old' machine-paced and restrictive ones and 'newer' lines with more operator responsibilities and generally more enriched jobs. Unfortunately the comparison was limited to that between the two situations and did not look at the 'improved' line before it was changed. The investigators used job satisfaction scores on work context variables and also took performance measures. Results in the direction expected, that the 'enriched' line was 'better', were found using the 'work itself' job satisfaction scores and sick leave absences. However, the interest is in how the authors explain problems or absence of effects for other variables. Personnel records were incomplete, other work environment factors confound comparisons, novelty of change must be allowed for, environmental and national economic changes can swamp any job design ones, and the job design change itself can disguise any measurement interpretations (e.g. data on errors and rejects).

In the second paper, Rahman focuses on health and stress as indicators of good or bad work designs in Bangladesh. This is an often forgotten issue in the usual search for attitudinal or performance measures (see references to Broadbent and to Warr in the later chapter from Martin). The author used Bengali versions of several well-known questionnaires and rating scales, comparing workers in self-paced repetitive and non-repetitive jobs, and consistently finding less stress and better physical and mental health amongst the latter. Despite apparently clear-cut results, even here the interpretation of these data requires much analysis, in this case in the light of high or low performance by the workers.

Finally, the paper from Venda of the USSR presents a new approach to ergonomics in work design. Basically he has taken a number of 'laws of co-adaptics' and suggests that these will aid the ergonomist to develop work redesign methodologies and to forecast the adaptation of people to machines and vice versa. This systems theory based approach is a nice contrast in philosophy to the more common psycho-social approaches used in the first two papers.

Chapter 22
Job redesign on the assembly line: some effects on job satisfaction and work behaviour

Jandre Šarić

SIZ za zapošljavanje Split, Žrtava fašizma 4, 58000, Split, Yugoslavia

Zora Raboteg-Šarić and Vladimir Takšić

Department of Psychology, Faculty of Science and Arts, Obala M. Tita 2, 57000 Zadar, Yugoslavia

This study investigates the effects of job redesign efforts in a sewing machine factory on job satisfaction, sickness absence, turnover, and quality of output. Two types of assembly lines were compared: 'old' lines that include machine-paced, monotonous and highly specific work and 'new' lines with wider variety of tasks, determination of work space, responsibility for quality control and improved job conditions.

The biographical and work attitude data were administered three times over a two year period to the samples of subjects who worked on different types of lines. The adapted version of Job Description Index was applied to give a measure of job satisfaction. Sickness absence, turnover, and data on quality of output over a three year period were obtained from company records.

The findings suggested that job redesign had some positive effects on work attitudes and behaviour. The work outcomes most strongly associated with job redesign are significantly higher job satisfaction (especially satisfaction with the work itself), and lower absenteeism due to sickness among the employees who worked on 'new' lines. Turnover rate was also lower. However, this difference was not significant. The relationship between job redesign and quality of output proved to be more complex. Workers with redesigned jobs accounted for more errors but less products rejected as waste.

Introduction

The primary goal of the classic scientific management approach was to design jobs for the most efficient production and to enhance organizational profit. More modern opinion, though, is that trends towards job simplification and specialization lead to boredom, job dissatisfaction, increased absenteeism, labour turnover and other inappropriate work behaviour.

Research attention has gradually shifted toward examination of the effects jobs can have on the people who perform them. There has been increased interest in designing jobs to make them generally more meaningful and challenging to employees. Protagonists of a motivational approach to job design emphasise the importance of enlarging and enriching the basic content of the job by adding dimensions such as variety, autonomy, feedback and control. Behind such a redesign proposal stands the argument that with a relatively small investment, this approach may have favourable effects on employees' satisfaction, productivity, turnover and absenteeism. Dowling (1977) has reported some job enrichment experiments in big companies which have been considered successful. Many other studies have also demonstrated that job redesign can have substantial impact on employee behaviour and attitudes.

However, some serious questions concerning the methodological rigour of a number of job enlargement studies have been raised, such as: a lack of appropriate control in these studies (Bishop and Hill, 1971), the consideration only of responses to questionnaires and interviews and not of actual behaviour of workers (Simonds and Orife, 1975), unintended and negative reactions of some workers to job enlargement (Hulin and Blood, 1968), and the absence of any systematic conceptual or theoretical basis for the experiments (Hackman and Lawler, 1971). Hackman and Oldham (1975) have presented a systematic, testable theory to explain the impact of motivating job characteristics on workers' attitudes and behaviour. Their job enrichment theory (a Job Characteristics Model) predicts that the enrichment results in high job satisfaction, high quality of work, and better attendance. Empirical results support the hypothesis that the presence of key job characteristics is strongly related to job satisfaction measures, but the relationship with the measures of productivity is rather weak.

The aim of this study was to evaluate the effects of job redesign on sewing machine assembly lines and to check the stability of effects over time. Job satisfaction, labour turnover, absenteeism, and the output quality of workers were compared between the standard existing production lines ('old') and production lines redesigned according to some job enrichment requirements ('new').

Method

Research settings and subjects

This investigation was undertaken in a large sewing machine factory of approximately 3000 employees. Job assignments on existing 30-operator assembly lines included highly specific, machine-paced and monotonous work operations. The analysis of efficiency of these lines had shown a relatively high percentage of waste, and high labour turnover and absenteeism rate. In order to eliminate the deficiencies of these lines some new assembly lines were constructed. The designers initiated changes in workplace layout, breaking big lines into smaller work posts (sitting in a circle so that communication was facilitated), and improving job conditions. Jobs were enriched by having employees set their own pace, and enlarged by introducing a wider variety of tasks and responsibility for quality control.

The subjects were all assembly workers who were present at the job on the days of investigation. The questionnaire survey was conducted three times over a two year period with different groups of subjects. The sample sizes were 59, 71 and 46 'old' lines workers and 35, 49 and 44 'new' lines workers. The average age for different samples ranged from 26 to 30 years, with average work experience from 6 to 9 years.

Measures and procedures

Job satisfaction and biographical data were collected by a specially designed questionnaire. Five aspects of job satisfaction (work, promotion, pay, supervisor and co-workers) were assessed by the use of an adapted version of the Job Description Index (Sabadin, 1978) which consisted of 105 items.

The biographical data collected included employees' age, sex, qualifications and work experience.

The questionnaire was administered anonymously to groups of assembly line workers on three occasions during a two year period, in October 1982, June 1984 and October 1984. The first occasion was 10 months after the job redesign changes had been introduced.

Additional data on voluntary turnover and sickness absence (total number of hours lost by the employee per year) were taken from the personnel files over a three year period (1982–4). Because of incomplete personnel records, sickness absence data were available only for workers who were employed in 1984. The company kept records on cumulative numbers of errors and numbers of products rejected as waste for a one year period. The records covering a period of two complete calendar years (1983–4) were available. These data were taken as measures of quality of output.

Results and discussion

Table 22.1. Parameters for job satisfaction (JDI) scores

Assembly line	Parameters	Job satisfaction (JDI) Work	Supervisor	Co-workers	Pay	Promotion
First survey						
OLD	M	20.22	30.73	51.08	4.11	11.11
(N = 59)	SD	8.58	22.74	18.48	4.58	6.65
NEW	M	33.17	47.20	50.94	7.65	12.48
(N = 35)	SD	10.89	20.96	20.86	6.93	7.76
	t	6.01§	2.05*	0.03	2.69§	0.87
Second survey						
OLD	M	23.25	46.63	47.93	6.70	11.42
(N = 71)	SD	11.66	19.91	22.52	6.80	8.00
NEW	M	30.73	39.98	48.14	6.57	9.90
(N = 49)	SD	12.91	22.48	20.52	6.40	7.90
	t	3.24§	1.67	0.05	0.11	1.03

Note: two tailed t-test for unrelated samples: * $p<0.05$, § $p<0.01$

The data on job satisfaction in the first two surveys are shown in Table 22.1. The results of the first showed significantly higher satisfaction with 'work', 'supervisor' and 'pay' among employees on the new assembly lines. The only significant difference in the second survey was found on the 'work' aspect of job satisfaction.

There may be a number of factors in the work environment (such as changes of supervisors or pay increases) that could influence the results on the other scales of JDI, besides the job redesign changes. Higher satisfaction with other aspects of the job in the first survey, but not the second, could also be explained by a 'novelty effect' which dissipated with time. In the third survey a 'work' subscale was the only one applied. The workers on redesigned assembly lines were again more satisfied with their jobs (M = 26.11, SD = 13.39, N = 46) than their colleagues on the old lines (M = 16.68, SD = 8.96, N = 44); this difference was also significant ($t = 3.94$, $p < .01$). Consistent differences in satisfaction with the work itself in the course of time suggest that job redesign had positive effects on work attitudes. Because changes in job design had primarily focused on job content, the 'work' subscale of JDI reflects them best.

Table 22.2. Sick leave (hours/year) parameters

One year periods	Old lines M	SD	N	New lines M	SD	N	*t*	*p*
1982	190.90	202.10	74	101.70	165.80	53	2.73	<.01
1983	133.97	153.56	95	63.80	87.00	59	3.41	<.01
1984	203.30	226.90	137	99.13	128.90	62	4.10	<.01

The analysis of sick leave is presented in Table 22.2. Significantly lower absenteeism due to sickness among the employees who worked on new lines was found for three one-year periods. This result could be seen as the side effect of improved work satisfaction and job conditions. Additional analyses of results showed that workers' qualifications, age or sex did not account for the difference in sick leave. This could be an additional indicator of positive effects of job redesign.

Labour turnover rates (percentage of workers who voluntarily left assembly line jobs) on the old assembly lines in 1982, 1983, and 1984 were 7.6%, 12.9%, and 11% respectively. Turnover rates on new lines were lower (3.65%, 8.6% and 3.5%) for the same period of time. However, these differences were not statistically significant. Turnover was generally lower than in previous years, probably because of the economic slowdown and restricted opportunities for employment. It may be that the effects of job redesign changes were undermined by such factors outside the organization.

The relationship between job redesign and quality of output proved to be more complex. Workers with redesigned jobs accounted for more errors in comparison with workers on old lines. However, while rejects existed on the old assembly lines, there were no rejects at all on the new lines during the period of study. Due to different methods of quality control and work organization on old and new assembly lines, these data are difficult to interpret.

In conclusion, the findings suggested that the work outcomes most strongly associated with job redesign were significantly higher job satisfaction and lower absenteeism due to sickness. Positive, though not significant, effects on output quality and turnover were also found. However, all these results must be treated with caution due to the absence of any base-line data prior to the work redesign changes.

References

Bishop, R.C. and Hill, J.W. (1971). Effects of job enlargement and job change on contiguous but nonmanipulated jobs as a function of workers' status. *Journal of Applied Psychology*, **55**, 175–181.

Dowling, W.F. (1977). Job redesign on the assembly line: farewell to blue-collar blues? In *Perspectives on Behavior in Organizations*, edited by J.R. Hackman, E.E. Lawler and L.P. Porter. (New York: McGraw-Hill), pp. 227–242.

Hackman, J.R. and Lawler, E.E. III (1971). Employee reactions to job characteristics. *Journal of Applied Psychology Monographs*, **55**, 259–286.

Hackman, J.R. and Oldham, G.R. (1975). Development of Job Diagnostic Survey. *Journal of Applied Psychology*, **60**, 159–170.

Hulin, C.L. and Blood, M.R. (1968). Job enlargement, individual differences and worker responses. *Psychological Bulletin*, **69**, 41–55.

Sabadin, A. (1978). *Analiza vprašalnika o zadovoljstvu pri delu*, unpublished Master of Science Thesis, (University of Ljubljana).
Simonds, R.H. and Orife, J.N. (1975). Worker behavior versus enrichment theory. *Administrative Science Quarterly*, **20**, 606–612.

Chapter 23
Stress and health in self-paced repetitive work

M. Rahman
Social and Applied Psychology Unit, University of Sheffield, Sheffield, UK

Abstract

This study is an attempt to investigate the pattern of stress and health in self-paced repetitive work. A sample of 150 self-paced workers was compared with a sample of 50 non-repetitive workers in terms of some psychological and physiological parameters. The findings of the study indicated that the self-paced workers experienced more stress and had poorer mental health than the non-repetitive workers.

Introduction

A large number of studies has already been carried out in industries of the developed countries to investigate the impact of repetitiveness and pacing on stress and health (eg. Cox, 1980; Salvendy, 1981). However a brief review of literature on repetitive and paced work indicates that reports of the impact of repetition and pacing on stress, performance and health are still equivocal. Although some studies have reported worker complaints of angina, headache, slight nervous disorders (Cooper and Marshall, 1976; Teiger and Laville, 1972), low-back disorders, and an adverse effect of repetitive work and dehumanizing environments on physical health (Shepard, 1971), it is not yet clear whether these problems result exclusively from repetition and pacing or from other factors such as workplace design (Beith, 1981).

Keeping these views in mind, the present study was designed to investigate the nature and pattern of stress and health in self-paced repetitive work in an industry of a developing country, viz., Bangladesh.

Method

Sample

A sample of 200 workers was randomly selected from a pool of about 1000 workers in different departments of a match factory in the city of Dhaka, Bangladesh. Out of these 200 workers, 150 were selected as undertaking self-paced repetitive work and the remaining 50 were taken as representative of non-repetitive work. The average ages of the self-paced repetitive and non-repetitive workers were 35 and 36 years respectively. The mean experience of the two groups was 14 and 16 years respectively, and mean years of education was 3 for each group.

Measures

Bengali versions of the following scales were used: Job Satisfaction Scale (Brayfield and Rothe, 1951), Stress and Arousal Check List (SACL) (King et al., 1983), General Health Questionnaire (GHQ-12) (Goldberg, 1972), Inventory of Subjective Health (ISH) (Dirken, 1967), Ratings of Perceived Exertion (RPE) (Borg, 1973) and Work Environment Questionnaire. In addition heart rate was measured in a selected group of subjects from the total pool of 200.

Procedure

Data were collected individually from the subjects on three consecutive days. Total time required for collection of data from each subject was 90 to 120 minutes. On the first day the Job Satisfaction Scale, Inventory of Subjective Health, GHQ-12, Work Environment Questionnaire and an interview schedule were administered. On the second day the SACL and RPE were administered on a pre and post work basis. Heart rates from a total of 61 repetitive and 21 non-repetitive workers were taken on a pre and post work basis. This measure was taken using a 77065 Digital Heart Rate Monitor which electronically records heart rate accurately, rapidly and conveniently.

Other personal information such as any physical disease, smoking habits, etc. were collected with the help of the interview schedule.

Results

Analysis of reported physical diseases showed that 53 per cent of self-paced repetitive against 38 per cent of the non–repetitive workers reported some sort of physical disease, but the difference was not significant (Table 23.1).

Table 23.1 Distribution of self-paced repetitive and non-repetitive workers according to reported physical disease

Groups	Physical Disease		Chi-square
	Yes	No	
Self-paced repetitive	79 (53%)	71 (47%)	3 ns
Non repetitive	19 (38%)	31 (62%)	

In the case of smoking it was found that 70 per cent of the self-paced repetitive group had this habit whereas only 42 per cent of the other group reported the same (Table 23.2).

Table 23.2 Distribution of self-paced repetitive and non-repetitive workers according to smoking habit

Groups	Smoking habit		Chi-square
	Yes	No	
Self-paced repetitive	105 (70%)	45 (30%)	12.59 $p<.01$
Non repetitive	21 (42%)	29 (58%)	

Additionally a significant trend was observed when the time of adopting the smoking habit was analysed. It was observed that 70 per cent of smokers of the repetitive group, against 48 per cent of the non-repetitive workers, developed the habit of smoking after starting their present work (Table 23.3).

Table 23.3. Distribution of smokers according to the time of developing the habit of smoking

Groups	Smoking Habit		Chi-square
	Before joining this work	After joining this work	
Self-paced repetitive	31 (30%)	74 (70%)	4.1 $p<.05$
Non repetitive	11 (52%)	10 (48%)	

Work environment

There were fifteen items regarding different aspects of work and the work place. The responses were analysed by applying chi-square tests. It can be seen from Figure 23.1 that the self-paced repetitive group perceived the environment more unfavourably than the non-repetitive group in terms of: the space needed for proper leg placement (62% against 30%, $\chi^2=15.16$, $p<.01$), repetitive physical effort required (93% against 72%, $\chi^2=15.48$, $p<.01$), long training required (48% against 20%, $\chi^2=12.39$, $p<.01$), the presence of harmful chemicals in the air of the room (60% against 20%, $\chi^2=23.63$, $p<.01$), the extent of social relationships possible during work (32% against 84%, $\chi^2=40.5$, $p<.01$) and the necessity for an additional short rest pause during work (63% against 34%, $\chi^2=12.81$, $p<.01$). On the other hand, non-repetitive workers perceived the work more unfavourably than the self-paced repetitive group in the case of nature of organisation of the work (60% against 78%, $\chi^2=5.05$, $p<.05$).

Stress and health

A series of t-tests was performed to determine the significance of the differences between the two groups across the measures taken in this study to assess stress and health (Table 23.4). It can be seen from Table 23.4 that the self-paced repetitive group differed from the non-repetitive group in terms of stress and health. Perceived effort expenditure of the repetitive group was significantly higher (both before and after work) than that of the non-repetitive group ($t=9.88$, $p<.01$). The groups did not differ in the case of stress and arousal before work, but after work, stress was significantly higher ($t=2.60$, $p<.01$) for the repetitive than the non-repetitive group, although arousal showed no significant difference. Self-paced repetitive workers were found to have lower job satisfaction (mean = 47.54) than the non-repetitive group (mean = 51.52). The GHQ scores ($t=6.66$, $p<.01$) were significantly higher (high score means better mental health) for the non-repetitive group than for the repetitive group (Table 23.4).

Similarly the inventory of subjective health score was much lower (indicating better health) for the non-repetitive group (mean = 11.46) than the self-paced repetitive group (mean = 16.96). In the case of job related performance i.e. absenteeism, the non-repetitive group were significantly better performers than the repetitive group ($t=3.99$, $p<.01$).

It can be seen from Table 23.5 that the groups did not differ in terms of heart rates before work but the heart rates after work of the self-paced repetitive group (mean = 92.13bpm) were significantly higher than for the other group (mean = 87bpm).

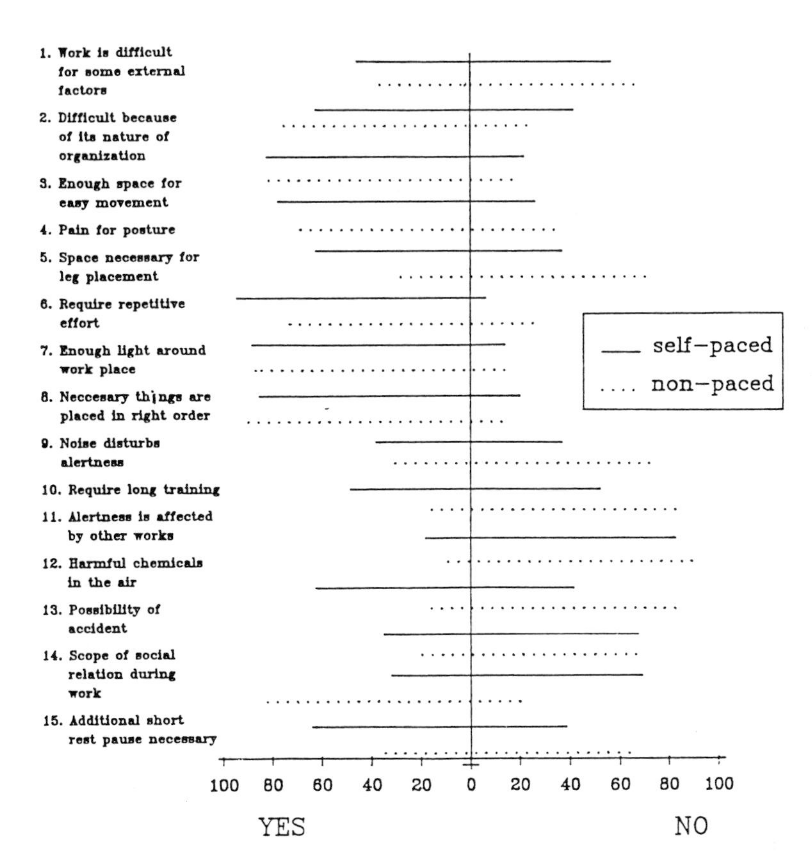

Figure 23.1. Percentages of responses of self-paced repetitive and non-repetitive workers to different aspects of work environment

Table 23.4. Mean and standard deviations of the study variables for self-paced repetitive and non-repetitive workers

Variables	Self-paced repetitive n = 150		Non-repetitive n = 50		*t* value
	Mean	sd	Mean	sd	
Before work RPE	13.24	1.40	11.50	.92	9.88§
After work RPE	17.37	1.95	15.40	1.32	8.20§
Before work stress	6.00	.63	6.34	1.02	1.61
After work stress	7.90	.86	7.04	.56	2.60*
Before work arousal	8.87	.87	8.76	.87	.45
After work arousal	6.63	.64	6.90	.97	1.28
Job satisfaction	47.54	5.21	51.52	4.27	5.45§
GHQ-12	14.19	4.30	18.19	4.01	6.06§
Subjective health	16.76	7.87	11.46	6.15	4.90§
Absenteeism	27.93	6.28	24.48	5.37	3.97§

*$p<.05$ §$p<.01$

Overall it appears that both psychological and physiological indices indicated higher stress levels and poorer mental and physical health of the self-paced repetitive workers than of the non-repetitive workers.

Table 23.5. Before and after work mean heart rates of self-paced repetitive and non-repetitive workers

Variables	Self-paced repetitive $n = 61$		Non-repetitive $n = 21$		t value
	Mean	sd	Mean	sd	
Before work	86.14	4.30	86.47	5.05	.26
After work	92.13	5.05	87.00	5.10	4.00§

§$p < 0.01$

Discussion

The findings of the present study consistently indicated that the non-repetitive workers experience less stress and possess better mental and physical health than do the self-paced repetitive workers. There are some studies (Eichar and Thompson, 1986; Karasek, 1979) which reported that people's control over their job and their occupational complexity (in terms of lack of repetition or routine) has a positive relation to job satisfaction. Recently, Sutton and Kahn (1986) also reported that control moderates the relations between perceived role stress, job satisfaction and psychological well-being. The present findings indicate that, although the self-paced repetitive workers have control over their work cycles, nevertheless it is possible that the repetitive nature of the job and the cognitive time tracking factor responsible for maintaining a high performance level during self-paced work put them under stress, which ultimately affects their health and well-being.

There are some studies (Manenica, 1977; Salvendy and Humphrey, 1979) which reported what Knight and Salvendy (1981) called surprising findings of lower sinus arrhythmia in self-paced tasks than in externally paced tasks. The findings indicated that, possibly, an internal pacing mechanism was responsible for maintaining the workers' pace during self-paced task performance. More recently, Rahman (1986) observed that high performance self-paced workers experience more stress and strain than low performance workers; the high performers maintain their performance level, however, at the cost of health and well-being.

The previous findings of lower sinus arrhythmia scores in self-paced tasks seem to support the present findings of higher stress and poorer physical and mental health status among the self-paced repetitive workers than the non-repetitive workers. From the Perceived Exertion Scale and Work Environment Questionnaires it seems that self-paced repetitive workers put more effort into maintaining their performance and they also perceived the job more unfavourably in terms of repetitive effort than did the non-repetitive workers. Thus, the perception of high energy expenditure may induce some sort of internal pacing mechanism in the self-paced workers.

If we consider that there is an internal pacing mechanism then it will indicate that self-paced work is not actually self-paced because it is highly unlikely that such a regular coordinated movement can be achieved by 'random activity' (Manenica, 1977). It also implies that it would be fruitful if future research efforts are directed to the understanding of behavioural processes (stress and strains are components) in these types of work rather than to compare them as such. The attempts may provide us with a sound basis to achieve better work design in these types of work.

In conclusion it can be said that the present study by and large indicates that work of a repetitive nature, even in self-paced work, can generate stress and affect the health and well-being of the workers. The findings are significant in the context of a developing country because a large number of workers are engaged in this type of work; job designers should take this into consideration when redesigning so that they can minimize the levels of stress and enhance health and well-being in this type of work.

References

Beith, B. (1981). Work repetition and pacing as a source of occupational stress. In *Machine Pacing and Occupational Stress*, edited by G. Salvendy and M.J. Smith, (London: Taylor and Francis), pp 197–202.

Borg, G. (1973). Perceived exertion: a note on history and methods. *Medicine and Science in Sports*, **5**, 90–93.

Brayfield, A.H. and Rothe, H.F. (1951). An index of job satisfaction. *Journal of Applied Psychology*, **35**, 307–311.

Cooper, C.L. and Marshall, J. (1976). Occupational sources of Stress: a review of the literature relating to coronary heart disease and mental health. *Journal of Occupational Psychology*, **49**, 11–28.

Cox, T. (1980). Repetitive work. In *Current Concerns in Occupational Stress*, edited by C.L. Cooper and P. Payne (Chichester: John Wiley).

Dirken, J.M. (1967). *Arbeiden Stress* (Groningen: Wolters).

Eichar, D.M. and Thompson, J.L.P. (1986). Alienation, occupational self-direction and worker consciousness. *Work and Occupation*, **13**, 47–65.

Goldberg, D.P. (1972). *The Detection of Psychiatric Illness by Questionnaire* (London: Oxford University Press).

Karasek, R.A., Jr. (1979). Job demands, decision latitudes and mental strain: implications for job redesign. *Administrative Sciences Quarterly*, **24**, 285–308.

King, M.G., Burrows, G.D. and Stanley, G.V. (1983). Measurement of stress and arousal: validation of the stress/arousal adjective checklist. *British Journal of Psychology*, **74**, 473–479.

Knight, J.L. and Salvendy, G. (1981). Effects of task feedback and stringency of external pacing on mental load and work performance. *Ergonomics*, **24** (10), 757–764.

Manenica, I. (1977). Comparison of some physiological indices during paced and unpaced work. *International Journal of Production Research*, **15**, 261–275.

Rahman, M. (1986). Performance, stress and strains in self-paced repetitive work. *Journal of Human Ergology*, **15**(2), 123–130.

Salvendy, G. and Humphrey, A.P. (1979). Effects of personality, perceptual difficulty and pacing of a task on productivity, job satisfaction and physiological stress. *Perceptual and Motor Skills*, **49**, 219–222.

Salvendy, G. (1981). Classification and characteristics of paced work. In *Machine Pacing and Occupational Stress*, edited by G. Salvendy and M.J. Smith (London: Taylor and Francis).

Shepard, J.M. (1971). *Automation and Alienation* (Cambridge, MA: MIT Press)

Sutton, R. and Kahn, R.L. (1986). Prediction, understanding and control as antidotes to organisational stress. In *Handbook of Organizational Behavior*, edited by J. Lorsch, (Englewood Cliffs: Prentice Hall).

Teiger, C. and Laville, A. (1972). The nature and variation of mental activity during repetitive tasks: an attempt to evaluate work load. *Le Travail Humain*, **35**(1). 99–116.

Chapter 24
Co-adaptics and transformatics in methodology and practice of work design

Valery F. Venda

Department of Learning, Research Institute for Higher Education, 103062, Moscow, USSR

Introduction

In addition to the traditional methodological base of ergonomics in work design the author proposes two new scientific directions: Co-adaptics and Transformatics. Co-adaptics, as a partial case, is a theory of mutual adaptation of people with machine and environment. In general Co-adaptics is a theory of mutual adaptation in all kinds of systems including living and non-living, ecological, technological and sociotechnical. Co-adaptics is based on the laws of mutual adaptation. The methodology of mutual adaptation of people, work, means and environment has been very effective in the practice of operators' work design in many Soviet industries—power plants, energy systems, chemistry, metallurgy, traffic control systems and others, designed with co-workers (Venda and Oschanin, 1962; and Mitkin, 1969; Venda, 1973). Important consequences of the Co-adaptics approach were a theory and practical methods of individual adaptation of information display to the operators (Venda, 1967, 1976, 1980c), of human–computer dialogues (Venda, 1976, 1977), and synthesis of adaptive, collective, computerized Hybrid Intelligence systems for solving the most complicated tasks in process control, forecasting of technological progress etc. (Venda, 1977, 1978).

Transformatics is a theory of dynamic processes during transformations of systems' structures. It is based on the law of transformations. The main consequences of Transformatics are the Transformation Learning Theory (Venda, 1980a, 1980b, 1983, 1984, 1985, 1986b), that was very fruitful for operators' professional training and adaptation to work places, and the general Transformation Theory of wave-like macroprocesses in nature, society and technological progress. The Transformation Theory describes dynamics of changing structures of control in human–machine–environment production systems and also of social structures like the 'perestroika' (reconstructing) that is under way in the USSR.

For application of Co-adaptics and Transformatics in the practice of work design the criteria and psychological factors of human mental work complexity were measured. The main criteria were probability of successful decision making or the number of decision tasks made during a fixed time period; productivity of work; stress and strains; time of decision making. The main psychological factors of intellectual work complexity were recognized as the parameters of a mental (conceptual) model of the control object and the actual task.

Co-adaptics: fundamentals and application to work design

For the further development of work design methodology it was necessary to find the laws concerning the fundamental bases of creating, adapting and the evolution of human–machine systems. Fundamental laws of ergonomics and psychology are needed which synthesize practical, experimental and theoretical experience of these sciences and also of biological evolution and general system theories.

Experience in ergonomic work design, consulting, research and teaching has convinced the author of the necessity and possibility of searching for fundamental ergonomic and psychological laws of existence, in the development and evolution of natural and artificial systems. The laws have to describe the processes of mutual adaptation as an essence of existing systems and the development of all living and artificial systems and then complex human–machine systems. That is the way to Co-adaptics.

The laws proposed below combine a base of Co-adaptics as a general theory of mutual adaptation (MA) and a system-evolutionary approach to natural and artificial systems, for progress in human engineering of prospective Hybrid Intelligence systems and other adaptive human–machine systems (HMS).

The First Law of Co-adaptics for any system: 'The existence and development of any system is a process of mutual adaptation between the system and environment and between inner components of the system'.

A *structure* $\bar{S}_i$ of system S_i is a representation of the regularity of the process of mutual adaptation of its components. A *strategy* $\underline{S}_i$ of the system S_i is a regularity of functioning of the structure $\bar{S}_i$ in the processes of mutual adaptation of the system with its environment. An efficiency Q_i of a system with structure-strategy $\underline{\bar{S}}_i$ is a measure of mutual adaptation of its components. The general form of a characterization curve of the system structure is a link between efficiency (Q) and a factor of efficiency and complexity (F) of the system, and is shown by an inverted U curve.

The Second Law of Co-adaptics is for living systems: 'The existence and development of a living system is a process of mutual forecasting multilevel adaptation (MFMA) between the system and environment and between inner components of the system'.

MFMA of components of a living system (LS) means that social (s), psychological (ψ), physiological (ζ), biological (β) and genetic (γ) levels of the system must adapt to each other in the concrete structure of the system. This is on the basis of prognostication of future process dynamics of the mutual adaptation of the system as a whole with its environment.

Co-adaptics in work design demands the reaching of appropriate mutual adaptation of all factors and characteristics of human work: social (F_s), psychological (F_ψ), physiological (F_ζ), biological (F_β) and genetic (F_γ). Work design starts with analysis of general and personal experience of work design for all factors (F_{ws}), $F_{w\psi}$, $F_{w\zeta}$, $F_{w\beta}$ and $F_{w\gamma}$ and criteria (Q_i). Then prospective computer and experimental models of work design are made which are investigated and developed with new experience and recommendations for work design methodology (Venda, 1971, 1973, 1986b).

The Third Law of Co-adaptics is oriented to artificial systems: 'Existence and development of an artificial system is a process of mutual forecasting multilevel adaptation of inner components of the system and of the system as a whole with its environment'.

MFMA in artificial systems is a materialization of such abilities of living systems comprising the artificial system.

The Fourth Law of Co-adaptics was formulated specially for human–machine systems: 'Existence and development of a human–machine system is a process of mutual forecasting multilevel adaptation between the system and environment, human and machine and between inner components of the human and the machine'.

MA means adaptation of a machine to a person in technical progress, human engineering and operative functioning and adaptation of the person to the machine in evolution, learning, professional selection, training and operation.

Forecasting adaptation (FA) of human to machine means forecasting by ergonomists of future dynamics of the machine and environment and its realization in decision making, MFMA of the operator's components, sensory and motor activity, and interaction with the machine. FA of machine to the person means the machine is the materialization and model of designers' and ergonomists' predictions of human–machine interaction in future situations. Decision making in HMS is a result of MFMA, of the strategy of the human operator (S_O), and of the *a priori* strategies of ergonomists and designers (S_A). The average measure of influence of designers on operators' activities is a correlation coefficient $r_{S_O S_A}$. If $r_{S_O S_A} \rightarrow 1$ all juridical responsibility for operator errors is passed to designers and ergonomists.

Multilevel adaptation means work design at all levels: s, ψ, ζ, β and γ. Multilevel adaptation of the working place means adaptation of different details and the whole machine to the operators on many levels: individual–operative, individual, group (typological), contingent and total. One of the important consequences of the laws of Co-adaptics is a theory of Hybrid Intelligence Systems.

The Fifth Law of Co-adaptics: 'Each living system may have a number of different structures $\bar{S}_i$ and strategies $\underline{S}_i$'.

Each of these strategies has its own characterization curve Q_i (F_j), where Q_i is the efficiency of $\underline{S}_i$, and F_j is the factor of mutual adaptation of the system with the environment.

There are three characterization curves of the human operator's work strategies S_a, S_b and S_c shown in Venda (1986a). This fifth law of Co-adaptics confirms that the law by Yerkes and Dodson is only a partial case of the law of structures and strategies.

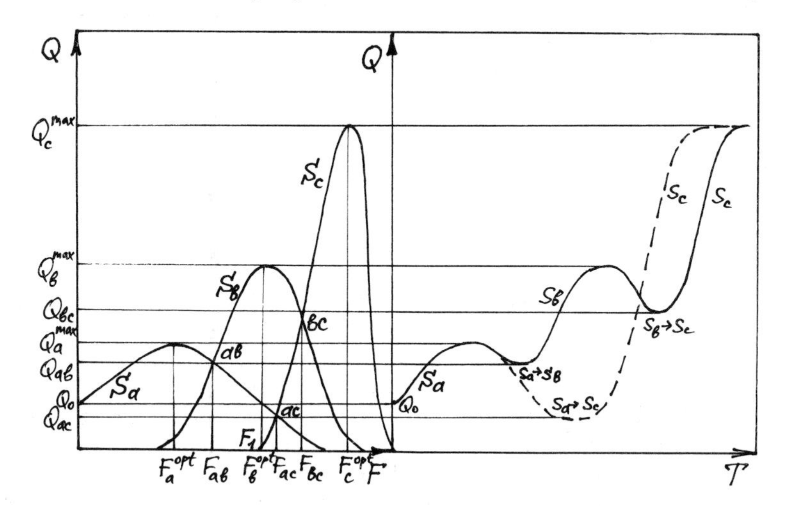

Figure 24.1. Characterization curves of different strategies S_a, S_b, S_c of human performance (left) and dynamics of its effectiveness during the learning process (right). Q is effectiveness, F is factor of effectiveness, T is time of learning.

Transformatics: fundamentals and application to work design

The Law of Transformations: 'The transformation of one structure $\bar{S}_i$ of a system to another $\bar{S}_{i+1}$ may be done through the common structures' state which is reflected by the point of intersection of the characterization curves of corresponding strategies $\underline{S}_i$ and $\underline{S}_{i+1}$'.

The Law of Transformations establishes that the structure-strategy S_a may transform to S_b over point ab if $F = F_{ab}$ and S_a may transform to S_c over point ac if $F = F_{ac}$ (see Figure 24.1). In the second case the decrease in efficiency of the system during a transformation will be more significant. One of the consequences of the Law is the transformation theory of learning and adaptation.

One can see in Figure 24.1 that if the human operator is an example of a system, one may watch different reactions with strategies S_a, S_b, S_c as $Q_a\ (F_1) \neq Q_b\ (F_1) \neq Q_c\ (F_1)$ with the same $F_\beta^{opt} = F_1$. This is one of the consequences of the Fifth Law. The Transformations S_a to S_b and S_c are possible over points ab and ac and corresponding levels Q_{ab} and Q_{ac}. This is one of the consequences of the Law.

It was found that for each structure and strategy of concrete human work there is a special characterization curve with its own optimal external conditions and maximal efficiency.

Methodology of work design as a synthesis of methodologies of Co-adaptics and Transformatics

The main idea in work design may be the mutual adaptation of people with work places (machine and environment), but the optimal condition for human performance depends on the concrete strategy of the performance. In Venda (1986a) one can see the experimental data from a compensatory tracking task with one to six simultaneously observed signals. According to the level of training and strategy used, the optimal number of signals might be between 1 and 5. So in work design the ergonomist has to reach the best mutual adaptation of human performance strategy and structure of the work place. As the most convenient graphic and computer model for work design, the Quadrigramme of mutual adaptation (Venda, 1988) is recommended.

Co-adaptics and Transformatics are important and useful parts of the general methodology and practice of work design.

References

Venda, V.F. (1967). Ergonomic research of information displays. *Ergonomics in Machine design*. (Geneva: International Labour Office).

Venda, V.F. (1971). Quelques perspectives de recherche dans la construction d'un systeme 'homme–automate'. *L'homme dans les systemes automatisés*, (Paris).

Venda, V.F. (1973). Ergonomie und Systemtechnik. *Ideen des exacten wissen*, **5** (Stuttgart).

Venda, V.F. (1976). Ergonomic problems of individual adaptation of operator's work means. Paper at the VI Congress of the IEA. *Ergonomics*, **3**.

Venda, V.F. (1977). Human factors: problems of adapting systems for the interaction of information to the individual: the theory of the Hybrid Intelligence. Opening address, *Proceeding of the 21st Annual Meeting of the Human Factors Society*.

Venda, V.F. (1978). Methodological principles of synthesis of the Hybrid Intelligence systems. *Proceedings of the International Conference on Cybernetics and Society*, (Tokyo).

Venda, V.F. (1980a). *Voies nouvelles pour une theorie de l'apprentissage*. Present et future de la psychologie du travail. EAP, (Paris).

Venda, V.F. (1980b). Psychological analysis of learning processes dynamics. *Proceedings of Finnish-Soviet symposium on Psychology of Work and Occupations*, (Helsinki).

Venda, V.F. (1980c). *Inginerska psychologia a synteza systemov zobrazovania informacii*: Praca, (Bratislava).

Venda, V.F. (1983). On the transformation learning theory. *Proceedings of the Third European Annual Conference on human decision making and manual control*, Risø National Laboratory, (Roskilde).

Venda, V.F. (1984). Today and perspectives of learning theory. *Soviet Psychology*, **XX**, 4.

Venda, V.F. (1985a). The transformations in learning processes and adaptation dynamics. *Proceedings of 1st Otaniemi symposium on work psychology*. Report No. 87.

Venda, V.F. (1986a). On Transformation learning theory. Behavioral science. *Journal of the Society for General System Research*, **31**, 1.

Venda, V.F. (1986b). On the laws of mutual adaptation in man–machine and other systems. *Trends in Ergonomics/Human factors* **III**, edited by W. Karwowski, (Amsterdam: North Holland/Elsevier Science Publ. B.V.)

Venda, V.F. (1988). The Quadrigramme of mutual adaptation as a new model of human activity. In *Ergonomics International 88: Proceedings 10th Congress of the IEA*, Sydney (London: Taylor & Francis).

Venda, V.F. and Oschanin, D.A. (1962). Wege zur Erhohung des arbeitseffekts des operateurs in Systemen 'Mensch-Automat'. *Probleme und ergebnisse der Psychologie, Sonderheft Ingenieurpsychologie*, (Berlin).

Venda, V.F., and Mitkin, A.A. (1969). Risultati di uno stadio obiettivo sull' attivita di un operatore. *La Scuola in Azione, Metanopoli* (San Donato Milanese—Giugno).

Part VII

Participative approaches to work design

Of all the criteria and keys to successful systems implementation, participation must be the most critical. Whether introducing new technical systems, new organisational structures or new job restructuring, initiatives which encourage the participation of those involved and affected will generally stand greatest chance of success. Participation, though, is an easy 'buzz word' to preach. Unfortunately, what purports to be such an approach often is a PR exercise or even an authoritarian system masquerading as an open, communicative approach. Moreover, as well as the often reported advantages of improved commitment, better information and solutions, and systematic spin-off benefits, participation can have many disadvantages and throw up several immense problems. These disadvantages, though, are most often seen when the philosophy and approach of participation are abused, and the problems are thrown up when the approach, methods and techniques employed are inadequate.

This section contains four very different approaches to the development and employment of participative systems. Daniellou, Kerguelen, Garrigou and Laville follow in the tradition of many different and exciting participative initiatives amongst French ergonomists. This particular study was to enable the intervention of workers in design for future work conditions at printing plants. Expecting workers to participate merely by consultation with drawings of future installations is not realistic, and so the investigators set-up a process of 'forecasting future activity'. Through this much wider focus the involvement of workers was made much more effective and also a comprehensive review of future needs could be made. The process for doing this is critical of course, a delicate matter requiring highly skilled and committed investigators. However, initial results are encouraging.

There is a strong Scandinavian tradition in the use of participation, or at least worker involvement, in solving work and other problems. The next two contributions are from Finland, but contrast greatly in their focus. Järvinen and Herranen take a participative approach to that issue of great concern in physical ergonomics, musculo-skeletal disorders. In two different factories a structured approach was taken to the identification of risk factors, development of improvements and their implementation. Standard questionnaire and observation methods were used, but in addition 'improvement suggestion' groups—comprising several levels and types of personnel—were set up. The group meetings proved remarkably productive for preliminary ideas, but less so in providing concrete solutions. Thus a number of suggestions are made regarding the process of organising, and the limits to benefits expected from, such a participative method.

Heiskanen reports an extensive—in time and remit—educational programme in which university staff arranged a programme for public administrators. The underlying themes were to bridge theory and practice in organisational change, division of labour, changes in management function and the introduction of computerised technology. Mixed results are reported but the author is sufficiently encouraged to suggest using the method to aid in organisation development.

Finally in this section, Zanders reports a Dutch participative study which took a standard analysis and decision technique, the Delphi Method, and applied it at an organisational level as an Interactive Delphi Methodology (IDM). Similarly to Daniellou and his colleagues, Zanders examines prediction, particularly involving groups of workers whose jobs will be affected by new technology. Within a structured process, participants are introduced to all the factors involved; then an interactive Delphi method is applied to questions of the time effects, anticipated workload and adoption or take-up of new technology.

These four papers represent just some of the participative processes available. Much remains to be done, not only in persuading organisations of the validity and utility of such processes, but in improving the methods and techniques used and in giving guidance on their appropriateness and use in different circumstances. As Zanders says, the proof of the pudding is in the eating.

Chapter 25

Taking future activity into account at the design stage: participative design in the printing industry

F. Daniellou and A. Garrigou

Laboratoire d'Ergonomie et de Neurophysiologie du Travail, Conservatoire National des Arts et Métiers, 41 rue Gay-Lussac, F75005 Paris

A. Kerguelen

Laboratoire de Psychologie du Travail, Ecole Pratique des Hautes Etudes, Paris.

and A. Laville

Laboratoire d'Ergonomie de l'Activité Professionnelle, Ecole Pratique des Hautes Etudes.

Origin of the demand

The ergonomics operations mentioned here concern the entire design of two newspaper printing plants under a French daily newspaper modernization programme. In order to face up to the challenge from European competitors, particularly in terms of speed and distribution targeting, two of the largest French press groups decided to modernize their printing plants.

In the press sector, specific data are available concerning the effects of work on health. These data, which came from epidemiological surveys (Lortie *et al.*, 1979, Teiger *et al.*, 1981) highlighted specific mortality and morbidity, particularly for the rotary printers. They were attributed to a combination of the following factors:

- working hours, in particular night work;
- the physical and chemical environment: noise (sound levels of 98 to 110 dB(A)), toxic substances (solvents and paper dust), and thermal conditions;
- a significant physical workload;
- a monitoring activity with high time constraint phases.

In view of these epidemiological data and the possibility during modernization to take account of working conditions, staff representatives negotiated with the management in order to have ergonomic studies integrated at the design stage of the new plants.

Presentation of the methodology

The guiding principle for these studies was an attempt to forecast the future activity of operators and their probable difficulties, in order to pinpoint the modifications which would be necessary for the working facilities. The methodological background is presented first and the course of the interventions carried out in the printing plants will be described afterwards.

The choice of a methodology aimed at forecasting future activity resulted from noting the limits of operations mainly based on the application of ergonomics standards. Although the latter were able to determine general targets in matters of workspace dimensions and the physical environment of workstations, they were unable to identify the difficulties which workers could encounter, particularly when dealing with production incidents.

The forecast of future activity is an extension of the work done in matters of activity analysis since Ombredane and Faverge (1955). This approach is characterized by a methodological choice: to analyse the work of the operators in real work situations (their actions, communication, reasonings), in order to identify the strategies which they apply, their technical and organizational determinants, and their consequences for workers' health and production.

Roughly speaking, from the historical viewpoint two periods relative to activity analysis could be distinguished: a period of diagnosis and a period of participation in the design of new production systems.

Initially, activity analyses mainly led to diagnoses and limited changes relative to existing situations:

- in mass production situations, activity analysis in particular highlighted industrial variability, including situations considered as repetitive (assembly line) and the cognitive activity used by the operator to face up to this (e.g. Teiger, 1978);
- analysis of the difficulties encountered at the time of computerization or automation of traditional situations. This work highlighted the fact that automation was quite often based on an over-simplified, or even inadequate view of the work done previously; designers frequently underestimated industrial variability and the control activity of workers who previously used to do the work in a traditional way. This often led to major difficulties for workers (learning difficulties, anxiety, accidents) and for companies (low utilisation rate of machines, late start-ups) (e.g. Wisner and Daniellou, 1984).

These are the difficulties which subsequently led companies and trade unions to ask ergonomists to take part in the entire design process for new production systems:

- participation in design of special facilities (computer or production control systems) (e.g. Wisner *et al.*, 1985);
- participation in the entire design of workshops or production plants (Cockerill-Sambre, 1984).

However, ergonomics methodology, based on assessment of the activity actually employed by the operator, has a problem when it comes to designing new systems, i.e. installations which do not exist as yet. Obviously, the activity in these installations cannot be observed. So, under these circumstances, how is it possible to assess the difficulties which an operator will encounter and adapt the systems accordingly? We are faced with the 'paradox of design ergonomics' (Pinsky and Theureau 1984).

Conditions for forecasting the future activity

'Forecasting the possible future activity' is aimed at solving this paradox. It is different from forecast methods used by designers and work organizers. 'Forecasting the possible future activity' is an attempt to describe the future use of human reasoning and the human body necessary in installations yet to be built both in 'normal' situations and also in preparation, incident, and maintenance situations.

This attempt is based on the interlinking of two phases:

- activity analysis in several existing comparable situations, which leads to identification of probable 'action situations' in the new installations;
- forecasting the future activity.

Analysis of the activity in existing situations

Several reference situations are sought: the workshop or the department to be modernized, but also situations where technologies similar to those planned in the project are used.

Activity analysis in these situations is directed by the data already available, concerning the future installations. This is aimed at highlighting:

- the difficulties encountered and the consequences for health and production;
- sources of variability (variation in raw materials, power networks, customer demands, wear of tools, sensitivity to weather conditions etc.);
- incidents liable to take place and critical periods;
- the activity employed by operators in order to prevent or correct these incidents, in particular the positive characteristics of the activity which should be safeguarded in future installations.

In certain cases, activity analysis in existing reference situations may take the form of lengthy and detailed observations. In other cases, due to the resource constraints relative to execution of the project, it may only be possible to make short observations, accompanied by in-depth interviews with the workers concerned.

Identification of probable 'action situations' in the new situations

On the basis of these various activity analyses, it is possible to pinpoint probable 'typical action situations' in the future installations.

A 'typical action situation' is defined by:

- the person(s) involved in an action;
- the targets aimed at (for example, preparing production or set-up, solving an incident, ensuring continuous operations, changing production parameters, cleaning the machine etc.);
- the constraints to which operators are subject (time constraints, quality demands, characteristics of materials or tools);
- factors liable to affect the internal condition of operators (night work, for instance).

For example, on a rotary press, there will be an 'action situation' for preparation of the run, start-up of the run, during the run, changes of edition, a paper tear, changing a reel and linking the new one, cleaning after a run and various maintenance operations. The 'probable action situations' in the future installation do not result immediately from analysis of the reference situations. Certain elements have to be ruled out in order to identify the transposable aspects of the situations observed.

'Action situations' can be highlighted and brought to the attention of designers before the specifications are drawn up. They will then serve as a basis for forecasting the future activity.

Forecasting the possible activity in different action situations

Designers' projects or proposals are put to the test of reconstitution of the future activity for the different 'action situations' noted. In certain cases, where a prototype is available, it is possible to carry out real ergonomics experiments. This method is particularly suited to interventions concerning a localized production system (software or machine). In other cases, especially that of large-scale assemblies, reconstitution can only be done on scale models or drawings.

The results of these reconstitutions are used to establish a prognosis relative to the work facilities planned and to ask for modifications.

A necessity: confrontation of several skills

The different stages presented previously are not produced by the ergonomist alone, or even through confrontation between the ergonomist and the technical managers of the project.

As we have underlined, a vital point involves identification of the sources of variability and the strategies used to solve them. Some elements of variability result from technical or commercial choices (introduction of variants, for example) and are known by project managers, but the skills relative to incidents and the ability to take them into account are usually not very developed in technical managers and designers. On the contrary, these are essential in production and maintenance operators, who have to deal with production incidents every day.

Due to this, task forces ('working groups') were organized as the basis for the ergonomic operation, in order to enable the examination of three types of skill:

- the skills of technical managers relative to the technologies adopted and the direction and control of the project as a whole;
- the skills of operators concerning dealing with variability and incidents, activities employed, and their consequences, particularly in terms of workload or health;
- ergonomic skills relative to knowledge about people at work, current activity analysis methods and forecasting the future activity. Ergonomists play an important part in organizing the work of the groups: in particular, they have to ensure that their work is actually based on the forecast of the future activity and is linked with the technical decisions relevant to the project.

The use of this methodology in two newspaper printing plants will now be presented.

Intervention in two newspaper printing plants

Structures set up

Working groups were set up according to the methodology presented previously. The groups comprised a project manager, the heads of the departments concerned, production and maintenance operators and ergonomists. In the first intervention, around 50 meetings were held over an 18-month period, each with a specific theme on the agenda. The same theme was repeated at different stages of the project. Working groups comprised a 'hard core' and other persons varied according to the subjects dealt with.

The function of the working groups was to pinpoint problems concerning future working and production conditions and to draw up recommendations intended for the various suppliers and the architect. The forecast reconstitutions took place at the level of the working groups.

A follow-up group was also created. This comprised trade union representatives, members of the company management and ergonomists. Its task was to define the main directions of the study, to follow up the progress of the working groups and see that their recommendations were taken into account.

Expert interventions

In certain fields (lighting, air conditioning, acoustics), the ergonomists drew up prior recommendations for the designers, with the possible help of other experts (acoustics and thermal experts).

Analysis of reference situations

There were two types of reference situation:

- first of all, activity analyses were carried out in the present printing plants of the groups concerned;
- secondly, visits were organized to European printing plants using technologies similar to those planned. These visits were made by the working groups mentioned previously. They led to observations of the activity.

Thanks to these two types of analysis it was possible to identify typical 'action situations' liable to be encountered in the future printing plants, which were used as a basis for forecasting the future activity.

Forecasting the possible future activity

Analysis of the possible future activity took place as follows. Each working group dealt with one aspect of the project. At the start of the meeting, the project leader presented the corresponding technical definition status. The group examined and completed the review of the 'action situations' concerned, established by the ergonomists on the basis of observations in reference situations.

For each of these situations, the group endeavoured to produce a detailed reconstitution of the corresponding activity using the several media made available: drawings, photographs, prototype software, models, screen copies, etc. The forecast was structured on the basis of a systematic examination of different operations, taking into account:

- requirements relative to tools and materials and their conditions of use;
- postures and efforts induced;
- the necessary communications and information;
- exposure to environmental factors;
- time constraints.

On the basis of this forecast, it was possible to highlight the probable difficulties for operators and for the company (supply defects, frequency or duration of incidents, newspaper quality, etc.) A 'prognosis' was then established, changes were requested and recommendations were drawn up.

At the end of each working group meeting, the ergonomists drafted a report containing remarks on the technical solutions, the forecast and also the recommendations made. In some cases, they took a stand that was different from or complementary to that of other members of the group.

Results

In the first printing plant, which is presently (January 1989) approaching completion, the forecast of the future activity led to numerous changes being made to the initial project. These concerned:

- the layout of premises and flows of raw material and staff. In particular, changes were introduced to allow for waste removal operations at the end of the run;
- exposure to physical and toxic environment factors: in some cases, the activity forecast led to suppression of the causes of exposure, particularly exposure to noise;
- production organization, in particular the paper supply circuit. The analysis showed that the inflexibility of the system initially planned made it difficult for operators to deal with incidents (paper tears, reel not accepted, etc.);
- physical efforts and work postures: analysis of probable interventions on the rotary press led to the installation of adjustable platforms on the machines and an elevator for access to the upper footbridges;
- the design of certain machines. For instance, the initial arrangement of the machine which addressed copies for subscribers led to a continuous output of bundles which the operators had to put in different post bags. This forced them to adopt high output rates and painful postures and made it difficult to meet postal regulations. A change was planned for the system to allow for the fact that the significant unit of work was the bag and not the bundle;
- software and information display consoles: the detailed forecasting of certain probable incidents led to changes in the structure of different software and information presentation consoles so that operators would be given the pertinent information and instructions for the actions which would have to be undertaken.

The different changes introduced will be assessed through an analysis of the activity or operators in the months following start-up.

Another type of result concerns the training of members of the working groups. It appeared that, for operators, their participation in the working groups constituted training for the new installations. The drawback resulting from the gap between training received in this way by members of the working group and that received by other operators could be compensated for by the systematic introduction in technical training programmes of future activity forecast exercises as a way of learning about the new installations.

The difficulties

Certain difficulties encountered were of a practical type and concerned obtaining agreement to and implementation of the changes requested, in view of the factors relative to deadlines, costs and work organization as defined in formal company agreements.

Other difficulties concerned the very approach of forecasting the future activity in the time context of the industrial project:

- forecasting the future activity was limited by the data and media available to the group at the time of the project: in certain cases, the inaccuracy or unreliability of suppliers' drawings, for example, led to wrong estimates of the necessary working spaces;
- the more the project advanced, the more precise the activity forecast became and the more difficult it became to make changes in it subsequently;
- as the project progressed, the interest in and relevance of ergonomic interventions became clearer for the various partners, their recommendations were increasingly appreciated, whilst at the same time the technical room for manoeuvre decreased;

- the actual working conditions not only depend on whether or not the recommendations are taken into account, but also on the care taken in the practical and detailed construction of the systems. It was therefore necessary to continue ergonomic intervention during the entire work phase and that of equipment installation.

Finally, certain difficulties were related to organization of the working groups and the need to reconcile a systematic approach with freedom of expression: a participant might bring up a new, significant element, which was not directly related to the subject being dealt with, but often it was neither possible nor advisable to put off the discussion which was to follow. The role of reports was vital in order to restructure the different contributions.

Conclusions

It appears that the approach proposed was particularly enlightening as regards the concept of 'workers' involvement in design'. A simple presentation of drawings of the future installation to the workers concerned rarely produces results for the following reasons:

- difficulties in understanding the drawings and the technology;
- focus on the technical aspects of the project to the detriment of work constraints; this makes it difficult for operators to represent their future activity;
- the difficulty workers have in assessing how their current knowledge is pertinent with regard to situations that are technically different.

By replacing this 'consultation' with a systematic approach of forecasting the future activity, the focus is shifted from the technical field to that of the future working conditions. This facilitates the confrontation of different types of skills, including those of operators, due to which the different partners can reach agreement concerning the changes envisaged.

References

Cockerill-Sambre (1984). *Prise en compte des facteurs humains dès la conception et l'installation de la coulée continue*, 4th Ergonomic Programme, CECA, (Luxembourg: EEC).

Lortie, M., Foret, J., Teiger, C. and Laville, A. (1979). Circadian rhythms and behaviour of permanent night workers, *International Archives of Occupational and Environmental Health*, **44**, 1–11.

Ombredane, A. and Faverge, J.M. (1955). *L'Analyse du Travail*, (Paris: P.U.F.).

Pinsky, L. and Theureau, J. (1984). Paradoxe de l'ergonomie de conception et logiciel informatique. *La Revue des Conditions de Travail*, **9**, 25–31.

Teiger, C. (1978). Regulation of activity: an analytical tool for studying workload in Perceptual Motor Tasks, *Ergonomics*, **21**, 3, 203–253.

Teiger, C., Laville, A., Lortie, M. (1981). Les travailleurs de nuit permanents, rythmes circadiens et mortalité, *Le Travail Humain*, **44**, 1, 71–92.

Wisner, A. and Daniellou, F. (1984). Operation rate of robotized systems: the contribution of ergonomic work analysis. In *Human Factors in Organizational Design and Management*, edited by H.W. Hendrick and O. Brown, (Amsterdam: Elsevier Science B.V.), pp. 461–465.

Wisner, A., Pavard, B., Pinsky, L. and Theureau, J. (1985). Place of work analysis in software design. In *Human Computer Interaction*, edited by G. Salvendy, (Amsterdam: Elsevier), pp. 147–156.

Acknowledgment

The following also took part in the operations presented: C. Teiger, CNRS and Conservatoire National des Arts et Métiers; F. Guérin and J. Maline, Agence Nationale pour l'Amélioration des Conditions de Travail.

Chapter 26
Case study: participative design in the prevention of musculoskeletal diseases at work

J. Järvinen and S. Herranen

Technical Research Centre of Finland, Occupational Safety Engineering Laboratory, P.O. Box 656, Tampere, Finland

Introduction

Neck and shoulder disorders are frequent amongst the workforce. These disorders can be found among, for example, packers, assembly operators, production line workers and office workers.

The combination of factors such as forceful and repetitive exertions, sustained or awkward postures and lack of rest may produce various degrees of discomfort and limitations of movement in the upper extremities. Factors that may affect the development of musculoskeletal diseases include also personal characteristics, psychological factors, time of exposure, experience and skills, etc. (Silta *et al.*, 1986).

In the prevention of musculoskeletal diseases it is necessary to try to affect several factors simultaneously; a shotgun approach therefore is used instead of a single-shot rifle when attempting to eliminate the possible causes (Silta *et al.*, 1986).

Musculoskeletal disease prevention strategies can be divided into five broad headings (Silverstein, 1987): engineering modifications (workstations, tools or parts), administrative controls (employee selection and restriction), organisation of work (enlargement, work groups, rotation, flowline reorganisation), training (work methods or ergonomic awareness) and personal protective equipment (pads, gloves, sleeves).

One approach for improving work and workplaces is to use participative design methods. Small groups have been used for recommending solutions to or solving problems related to productivity and/or quality within the framework of Quality Circle programs (Sen, 1987). Workers' participation in work and workplace design (or redesign) is necessary, because no one knows the workplace better than the people using it daily. However, in addition, the experts in different disciplines such as industrial engineering, machinery design and ergonomics should be involved.

This paper describes an attempt to improve the ergonomics of work and workplaces at two industrial enterprises. The purpose was to utilize the data, knowledge and ideas that already existed in different quarters at the enterprises.

The cases described here are included in a sub-project of the Finnish Work Environment Fund's research programme 'Musculoskeletal Disease Prevention at Work'.

The aim of the study was to determine what means of technical planning can be used to improve the work environment and materials handling. One function of the project is to contribute to the technical improvements in participating enterprises. These are a tyre factory and a box-making factory, which were selected because of the high rates of neck and shoulder disorders prevalent amongst their workforces.

In the tyre factory the study was concentrated on the tyre assembly department. Tyre assembly work includes fast handling, lifting and moving of tyre components in awkward postures. The weights of the components vary from 0.3–12 kg.

In the box-making factory cardboard is printed, punched, folded, glued and packed. The study was concentrated on the punching and glueing machine departments. The work at glueing machines is highly repetitive with machine paced loading, receiving and packing. The work at punching machines during the drive mostly concerns control and disturbance elimination, but their machine setting work requires working in awkward postures for long periods.

Identifying risk factors

Inquiry and interviews

A questionnaire concerning work and equipment was used to identify the problems connected with work. The questionnaire included 18 open questions, which dealt with job satisfaction, meaning of work, improvements and modifications made at the workplace, improvement suggestions and willingness to participate in the study.

The questionnaire was addressed to 200 workers (55 per cent responded) of the box-making factory and a cardboard mill, and to 208 workers (70 per cent responded) of the tyre factory and a footwear factory.

In both factories neck-shoulder and back problems were common. Workers were quite satisfied with their jobs. The inquiry revealed that some engineering modifications for improving work and the work environment had been made in both factories, but the workers were not involved in planning the changes. In the tyre factory the changes improved safety and eased the work, but also increased production standards (and repetitiveness).

The problems at workstations were associated with awkward postures, forceful and repetitive motions and defective tools.

Following this initial inquiry a group of workers and their workstations in the tyre factory and in the box-making factory was selected for a deeper examination. In the tyre factory a total of nine workers from workstations in the car tyre assembly department was selected. In the cardboard factory a total of 10 workers at the punching and glueing machines was selected.

The workers selected were interviewed concerning environmental and work related factors that affected them. The purpose was to extend the information received from the inquiry. The subjects were interviewed for about one hour during working hours.

More detailed information about the problems connected with subjects' work and workplaces was gained and the subjects also made additional suggestions for improvements during the interviews.

Job observation

The subjects' jobs were observed systematically with the aid of a form drawn up for this purpose. The form included space for basic job information, layout dimensions, environmental conditions, social contacts, flow of the work, data for estimating load on the lowback, ability of the workers to control their own work, feedback, pace, notes and suggestions and simplified OWAS records (OWAS is a system for analysing working postures). Photographs were taken of the workstations and work situations.

During the job observations the workers were asked again to propose improvements for their work and workstations.

Making improvement suggestions

To produce more suggestions for redesign of the work and workstations, improvement suggestion groups were founded in both participating enterprises. Both groups consisted of a core of permanent members with a worker from each individual workstation being co-opted as an ad hoc member.

Improvement suggestion groups in the box-making factory were composed of:

- co-opted worker
- occupational nurse
- personnel department representative
- researchers (engineer, occupational therapist and physician)

Improvement suggestion groups in the tyre factory were composed of:

- co-opted worker
- physiotherapist
- safety representative of employees
- chief engineer of assembly department
- researchers

Before each meeting the group members were given the summary of problems identified by questionnaire, by interviews and by job analysis. The members were encouraged to consider and record their improvement suggestions prior to meetings.

In the meetings the problems were discussed and suggestions were offered and written down. The aim of the meetings was also to identify new problems. The problems were discussed with the aid of slides taken of the workstations and work situations.

Improvement suggestions

The group meetings resulted in many improvement suggestions. The group at the box-making factory discussed eight workstations, and fifty-one improvement suggestions were made. The group at the tyre factory made fifty-seven suggestions concerning six workstations.

The distribution of suggestions in the box-making factory was:

– engineering modifications	37
– instructions	12
– work organization	2

Suggestion distribution in the tyre factory was:

– engineering modifications	39
– instructions	8

- physiotherapy counselling on job methods 5
- work organization 3
- personal protective equipment 2

Installation and monitoring

Implementation of improvements

The suggestions were presented to the production managers at the box-making factory. At the tyre factory the managers themselves attended the meetings.

The implementation of simple and inexpensive improvements was started immediately after each meeting.

Follow-up

Information about implementation, and costs and effects of the improvements was obtained during the follow-up phase. The effects were determined by interviewing the relevant personnel and by the same methods as in the problem identification phase.

A form was desiqued for the follow-up. It contained columns for individual suggestions, implementation decisions (yes/no), a schedule for implementation or reasons for delay or non-implementation, likely cost of intervention and possible effects of intervention.

About six months after the suggestion process, one third of the suggestions had been implemented in both factories. In the box-making factory the changes made were engineering modifications and provision of instructions. In the tyre factory counselling on job methods was introduced in addition to engineering modifications and instruction provision.

Examples of interventions implemented at the box-making factory are:

- limiting the height of punched cardboard stacks on pallets
- installing an inclined plane under the punching press waste disposal; cuttings slide down the plane from under the machine which removes the need to stoop to pull out the cuttings
- development of the jig for preparing the punching forms

The box-making factory will move to a new plant within one year and there is no intention of implementing more changes before moving to the new facility.

Examples of interventions implemented at the tyre factory are:

- redesign of the belt cutting knife and of the cutting motion required
- relocation of the counter (which is pushed manually after every work cycle)
- development of a prototype fixture for easing the tyre on and off the inflater

The schedule for implementation is late, because of the factory's limited design resources. The designers are busy with day-to-day problems.

Conclusion

This kind of participative approach for improving the ergonomics of work and workplaces seems to be effective as far as making suggestions for improvements is concerned.

It is important that the representatives of employees and also of the production organization take part in the meetings, together with people knowledgeable in ergonomics. The users of the workstations know the job best, and they are valuable in making and evaluating the improvement suggestions. Active participation by the users also reduces the resistance to change.

The acceptance of suggestions needs the executive production managers to understand the problems and the benefits of improvements. The production managers may have practical ideas for solving the problems identified or expressed in the group meetings and they have the power to implement the improvements. If the suggestions are presented only in the form of a summary the suggestions may appear irrelevant and be difficult to understand.

It is difficult to generate concrete solutions using this kind of group method. However, the groups produce many ideas which can be developed further. In addition to production engineers, machinery engineers and/or maintenance engineers should also be represented at the group meetings. By attending the group meetings they would get the background information on the problems and the basis required for more detailed design of the improvements.

It must be noted that normally the changes are made to increase productivity. Changes that ease, simplify or speed up the work will probably also increase repetitive loading and this may make it difficult to get workers to propose their ideas for developing their work.

Another sub-project of the research programme makes an attempt to implement the improvements generated for a larger group and will compare the impacts with a control group (no interventions) and with a 'neck school' group with no technical interventions.

References

Sen, T.K. (1987). Participative group techniques. In *Handbook of Human Factors*, edited by G. Salvendy, (New York: John Wiley & Sons), pp. 453–469.

Silta, J., Heikkilä, S. and Kuorinka, I. (1986). *Ergonomia toistotyössä-rasitussairauksien ehkäisy* (Porvoo: WSOY).

Silverstein, B.A. (1987). Evaluation of interventions for control of cumulative trauma disorders. In *Ergonomic Interventions to Prevent Musculoskeletal Injuries in Industry*, (Chelsea, MI: Lewis Publishers).

Chapter 27
Influencing organizational change: evaluation of a university-level educational programme

Tuula Aulikki Heiskanen
Department of Public Health, University of Tampere, Finland

Introduction

This paper describes a university-level educational programme for personnel directly involved in the planning and execution of organizational changes in state administration.

In Finland there have recently been major changes in the field of public administration. As yet not much is known about how these changes are going to affect the position and working conditions of the employees concerned, but in any event the implications are bound to be significant. The introduction of profit-oriented management and the service ideology within state administration has affected all levels from central administration down to the local level, and brought fundamental organizational rearrangements and innovations.

The educational challenge in this situation was to inspire a dialogue between theory and practice and to provide the participants with the basic tools they need for a critical scrutiny of their own activities. Ultimately, the aim was to encourage and help the participants to analyse the changes occurring in state administration and in its philosophy—or as Argyris and his colleagues (Argyris and Schön, 1975; Argyris, 1982) would have it, in the espoused theories—and to look at their own concrete problems from a new, broader perspective.

Methodologies

A university-level Institute for Extension Studies organised an educational programme on a multi-disciplinary basis. The focal themes of the programme were (1) the change of working life and the production of related knowledge, (2) the planning and development of work, (3) work organizations, their development and management, and (4) the social, psychological and health consequences of work. The programme was carried out within the space of one year. It was divided into five separate courses and included on-the-job assignments. The teaching methods adopted were essentially based on the model recommended by Kolb (1984), where the circle of experiential learning comprises four modes of learning: concrete experience, reflective observation, abstract conceptualisation and active experimentation. In accordance with Kolb's idea of concrete experience, the participants were encouraged to discuss the problems they encountered at their own workplaces; reflective observation was encouraged by arranging group discussions in which practical problems were considered from many different angles; abstract conceptualisation was encouraged through lectures and literature on theoretical concepts for analysing the problems; and finally, active experimentation was encouraged by on-the-job project assignments, where the participants had the opportunity to test in practice their new ideas.

A total of twenty-four public administrators representing central and local levels of administration took part: planners, educators, staff of personnel administration and other experts. One of the participants dropped out for personal reasons. In terms of background education, the majority had an academic degree and five had a college degree. A large number of disciplines were represented. Nine of the participants were men, fourteen were women.

The participants identified various specific pressures towards change, including organizational innovations aimed at changing the present division of labour (such as between central and local levels), changes in management ideology (resulting from the increased tendency to restructure public utilities to run on a commercial basis) and the introduction of automated data processing (ADP). Their project assignments—empirical research, studies of documents, literature reviews, etc.—were more or less directly related to these changes.

The programme was evaluated separately by the teaching staff, the students themselves, and by outsiders, who examined the study reports and other projects produced by the students. One year after the completion of the programme, a follow-up questionnaire was mailed to all participants. The present analysis of the programme is concerned especially with the following points:

1. the learning process and the participants' individual ways of capitalising on what they have learned;
2. their understanding of their own tasks at work and the changes in their work strategies; and
3. their chances of putting into effect the ideas generated in the course of the programme.

On the basis of the follow-up questionnaire, this paper also attempts to establish how, one year later, the education is reflected in the participants' intellectual pursuits and in their work strategies, and to find out what kind of development projects it has inspired.

Results

The fruitfulness of the dialogue between theory and practice depended essentially on the expectations of each participant with regard to the practical versus theoretical orientation of the programme. These expectations partly reflect individual study orientations, partly the participants' interpretations of the pressures in their respective organizational positions. The case analysis identified different types of learners with differing expectations of education and differing relative weights of Kolb's modes of learning. The practical developer gave priority to immediate concrete experiences. The one who applied reflectively also attached importance to other people's experiences and literary knowledge. The one who aimed for civilisation and sought new perspectives found theoretical knowledge most inspiring, even when this was of no immediate practical relevance. Finally, the one who strove for theoretically understood practice was particularly receptive both to the experiences of other participants and to abstract knowledge. During the programme it would have been important for the participants to stand back from their practical work-related problems, but this was quite obviously hampered by their expectations of speedy results, particularly in the case of the practical developer.

Given the programme's goal of producing both practical and theoretical results, the evaluation of the programme was also concerned with both these aspects (see Table 27.1). The participants felt that they had acquired a better knowledge of the field and were also better equipped to analyse its problems from a broader perspective. Some, however, complained that the specific problems of state administration had not received sufficient attention. Most of the participants also had new ideas for the development of their own tasks, either through a new approach or through practical rearrangements. None the less the relationship between theory and practice seemed to remain problematic: the programme offered no solution for linking the two, and this apparently made some participants sceptical with regard to the applicability of theoretical knowledge in general.

Table 27.1. Cognitive and activity-related changes inspired by the educational programme. (n = 23)

Development during the programme (response rate 96%)	
Cognitive	
Increased theoretical understanding of practical phenomena	9
Broadened perspectives	15
Increased knowledge of recent research results and trends	21
Increased knowledge of changes in other organizations and sectors of state administration	10
New ideas of reorganizing one's own work tasks	18
Development after the programme (response rate 91%)	
Cognitive	
Intends to follow developments in the field by reading journals, etc.	13
Intends to continue post-graduate studies	4
Re-focused interest in new fields	10
Increased scrutiny of own decisions on the job and weighing of different alternatives	18
Activity-related	
Has continued work on project theme	7
Produced publications related to the project's theme	6
Launched/participated in study circles at office	3
Organized education in the subjects covered in the course	5
Applied ideas generated by the education through committees	7
Launched/taken part in participatory research	2
Worked on major organization development projects	2

A large number of participants reported that they had continued to pursue their studies after the completion of the programme. They had tried to keep abreast of recent developments in the field by reading books and journals, even though they complained that there was not enough time for this. A few even planned to resume their post-graduate studies. The vast majority had attempted to put their new ideas into effect by arranging training schemes, setting up study circles, referring their ideas to committees, organizing participatory research, and working on major organization development projects.

Even during the programme many participants were able to form some picture of what their own work community thought about their new ideas; while working in their project they had to enter into a dialogue with that community. The expectations set on the part of the work community on the relevance and value of the project varied from a neutral, reserved attitude to positive support and interest. The main obstacles to the implementation of these ideas were felt to lie in time pressure, routines, lack of knowledge, adverse attitudes, and an established organizational culture. The participants realised that development work was a lengthy process, which required an unprejudiced attitude and also a tolerance of slowness. The new ideas should be widely spread, and the new objectives must also be adopted by the management before any real progress can be made.

Discussion

One of the goals of the programme was to encourage a conscious, critical reflection of practical work problems, but at the same time to help the participants to take a different perspective from the traditional ways of interpreting these problems. Most of the participants had some difficulty in adjusting themselves to the method of education. This of course was no surprise: it is always a difficult process to throw overboard existing thought patterns. Argyris (1982), for example, has shown how arduous a process it is to make people change their theories-in-use.

The programme provided the participants with some basic tools for analysing their work and for developing their own organization. Their expectations of finding ready-made solutions slowly gave way in the course of the programme, and gradually the participants adopted a more active orientation to studying. Scholars who have worked on the development of participatory research method (e.g. Gustavsen and Engelstad, 1986; Elden, 1983) have tended to be quite sceptical about the applicability of structured, general theories for the solution of problems at the workplace level. In place of such theories, they have offered so-called local theories, which are context-bound, in that they offer explanations for specific situations, and which are developed together with the people involved in those situations. The development of local theories has to start from the collective learning process. This process is enhanced by the kind of study circles, training sessions and research activities which were initiated by the participants at their places of work, in that they provide the tools for critical and analytical scrutiny of concrete problems.

The results suggest that this type of educational intervention may prove very useful for purposes of organization development. Although the method applied has the potential to produce good learning results, it is important to pay serious attention to the factors which may undermine this process, notably those that are related to individual study orientations and the boundary conditions set by the organizations. The educational system may select its students from amongst those who are in a position to carry into effect the things they learn in the course of their training, but it has no influence on the organizations themselves, on the structures and cultural elements which hamper the testing and introduction of new ideas. It seems that if more than one person is enrolled from the same organization, this will have a positive impact on the introduction of organizational innovations.

References

Argyris, C. (1982). *Reasoning, learning and action. Individual and organizational* (San Francisco: Jossey-Bass).

Argyris, C. and Schön, D.A. (1975). *Theory in practice. Increasing professional effectiveness* (San Francisco: Jossey-Bass).

Elden, M. (1983). Democratization and participative research in developing local theory. *Journal of Occupational Behaviour*, **4**, 21–23.
Gustavsen, B. and Engelstad, P.H. (1986). The design of conferences and the evolving role of democratic dialogue in changing working life. *Human Relations*, **39**, 110–116.
Kolb, D.A. (1984). *Experiential learning. Experience as the source of learning and development* (Englewood Cliffs: Prentice Hall).

Chapter 28

Participative work design. A systematic approach by the Interactive Delphi Methodology (IDM)

Harry L. Zanders

Tilburg University, and CPI-Foundation: Center for Productivity Research on Information-Work, Tilburg, The Netherlands

Introduction

The relationship between technology and employment has been discussed intensively during the last few years. Since the innovative introduction of information technologies in organizations it can be seen particularly that questions are raised about the effects of these technologies on the quantity and on the quality of work. Many studies are available on that topic, which give very different results and prognoses. It is clear that there is no consensus about the possible effects of the technology on the level of employment and the quality of work in the future. Optimistic and pessimistic scenarios are available. Sometimes we even see that the same empirical data are interpreted in opposite ways. One of the reasons for this lack of consensus is the variety of methodological approaches used in the employment studies.

In the literature several strategies and techniques are given that can be used to analyze the effects of automation on employment. Most of the methods are based on time series, with employment data that are extrapolated via regression techniques. A smaller category of methods uses the Delphi approach as an analytical strategy.

It is not so difficult to indicate some problematic points in both approaches, as Hunt and Hunt (1986) demonstrated. This article, however, will present some considerable advantages that the Delphi approach has to offer when it is applied not at the macro-level, but at a lower level, i.e. in organizations. This micro-approach is especially adapted to involve employees in the process of organizational change when new information technologies are introduced. The approach is named the IDM approach, where IDM means Interactive Delphi Methodology.

The 'Georgia Tech.' methodology

In 1985 a team of researchers at the Georgia Institute of Technology in Atlanta USA published the study: 'The Impact of Office Automation on Clerical Employment, 1985–2000. Forecasting Techniques and Plausible Futures in Banking and Insurance' (Roessner, *et al.*, 1985). The researchers developed a new methodology to estimate the effects of automation on the quantity of employment in specific sectors. The core of the procedure was a Delphi strategy named EFTE, which stands for: Estimate, Feedback, Talk, Estimate. The estimates are based on the predictions of experts, selected by the researchers for their expertise on information technology and on the relevant economic sectors. The experts give (anonymously) their first predictions and these are presented to the total group of experts (feedback). After the first results have been discussed a second round of predictions takes place. This is the classic iterative strategy of the Delphi approach. What is typical of the EFTE method is that the experts are operating in interactive group sessions.

Another characteristic of the Georgia Tech. approach is the (sub)division of the jobs studied into six dimensions. These dimensions are:

1 information intake,
2 information processing,
3 database management,
4 information production and transformation,
5 information dissemination,
6 interactive communication.

For all the six dimensions information technologies are available that will affect the quantity and quality of the work that is actually done. The effects of the information technologies on the six dimensions have to be estimated. The third characteristic of the Georgia Tech. approach is that estimates or predictions are given for the factors: time-saving, workload and adoption. Time-saving estimates give the reduction in working hours that is expected when new information technologies are implemented in organizations; workload estimates give a prognosis of the expected amount of work in specific jobs in the future; adoption prognoses give the expected percentage of employees who will adopt the new information technologies in their work.

During several EFTE-rounds estimates are given and discussed by the experts for the three factors—time saving, workload and adoption—on the six dimensions already mentioned. The final results are combined to get an overall estimate of the effects of information technologies on the quantity of employment.

Interactive Delphi Methodology: the IDM approach

In 1987 the Georgia Tech. approach was tested in a pilot project studying the employment effects of office automation in the non-commercial service sector or public sector in The Netherlands. The project was sponsored by the Dutch Organization for Strategic Labour Market Research: OSA (Organisatie voor Strategisch Arbeidsmarktonderzoek). The CPI-Foundation (Stichting CPI: Centrum Produktiviteitsonderzoek Informatie-arbeid) conducted the study in cooperation with Tilburg University. The results were published in a report by Zanders and Willems (1987)–Automation in the Non-Profit Service Sector, Consequences for Employment. The expected effects of automation were studied for three occupational groups: managers, professionals/staff and secretaries. Table 28.1 summarizes the main quantitative conclusions for estimated changes in employment:

Table 28.1. Estimated effects of office automation on employment

Employment	Managers	Professionals/staff	Secretaries
in 1990	2.5%	4.9%	9.1%
in 2000	3.4%	6.6%	12.9%

Actors as experts

The central experience in the Dutch pilot study has been that the EFTE-procedure opens up new ways to involve employees in the process of automation and innovation. The Georgia Tech. approach can be used if the role and the contribution of the experts in the method is changed. In the original application of the methodology the experts were persons with extensive professional knowledge of the possible effects of automation. The Dutch study prefers to use as 'experts' the people that will be directly affected by the effects of the new technologies in organizations.

Employee involvement and user participation are central issues in the discussions about the effective implementation of new technologies in organizations. Several methods are available to promote and to stimulate an active participation of the potential users of information technologies. A well-known and internationally widely used strategy is the ETHICS method, developed by Enid Mumford (1979), ETHICS being an acronym of Effective Technical and Human Implementation of Computer Systems. In The Netherlands such researchers as de Sitter (1982 and 1986) and van Eijnatten (1985) have presented participative strategies based on the socio-technical systems approach. An example is the STTA methodology: Socio-Technical Task Analysis. One of the problematic points of these methods is that they are rather complex and cumbersome in their application.

The IDM approach offers a direct feedback of the results to the participants. In one day a group of about 15 to 20 participants can be involved in an interactive process of discussions and presentations about the problems and potentials of new information technologies for their tasks and jobs. This can be done on the basis of a computer program, constructed to process the different steps in the Delphi methodology on a personal computer. Facts and especially opinions and speculations about the effects of automation are the input for the program. The main point is that there is a direct feedback of the individual estimates to the total group, so that the iterative Delphi procedure can be employed without delay for both processing and reporting.

Application of the method: eight steps

A general overview of the aims and the content of the IDM strategy has been given above. Now the concrete steps that have to be followed when the method is applied in an organization will be presented.

Before proceeding to the eight steps the following general comments must be given. The IDM approach is meant to assess the effects of technological systems on functions or jobs in organizations. The main reason for applying the method is the involvement of employees in the process of change that takes place when new technologies are introduced in organizations. Therefore the people invited to participate are those whose jobs will be affected by the technology. These employees are the actors in the IDM approach. Group sessions are organized where the possible and expected effects are discussed by the participants along the lines or steps of the IDM approach. The maximum size of the group in a session is about 15 to 20 persons. With more people in one group it can be difficult to manage the session and to structure the discussions. For the minimum size there is strictly no limitation, but with less than five participants the application of the method is hardly relevant

The eight steps of the method follow.

1 Selection of function(s) and participants

The first thing that has to be done is the selection of the functions or jobs that will be analyzed in a specific session. One single function can be chosen, but it is also possible to select several functions or jobs. In that case the best way to operate is to select jobs with comparable tasks or elements. That is not a question of principle but is in the first place a practical matter. The reason is that people will discuss possible effects of the technology on the selected jobs. When these jobs are rather similar and comparable it is easier to speak in the group, and the people involved in the application of the method will use the same frame of reference. If however the functions to be analyzed are very different then it is better to organize separate sessions, so that the different functions can be handled in sequence.

Some examples of functions that have been analyzed are: secretaries, middle managers in the banking sector, computer operators, staff employees with the same professional background (e.g. accounting).

When the function or some comparable functions have been selected, the employees who fulfil these functions are invited to participate in the program. If there are too many candidates two possibilities are open. The first is that a selection is made among the candidates to reduce the group to between fifteen and twenty persons. The second possibility is to organize more sessions so that all the candidates can participate in the process of involvement using the IDM approach.

2 Presentation of technological trends

During the IDM sessions people discuss the expected effects of technological developments on their work and their job. An important and necessary condition for these discussions is that all the participants have a sufficient view of the technologies that will be available and used in their organization in the near future, particularly the technology that the organization will introduce and implement in the functions or jobs being discussed in the sessions. That overview of the technological trends has to be presented to the participants before a session starts. The best moment to offer the information to the candidates is just one or two days before the IDM sessions. The presentation can be given by the information technology experts of the organization, on the basis of documents about the strategy for information technologies and information resources.

Not all organizations have such documents or even a strategy, and if documents are available they sometimes contain rather general statements with little detail about systems and techniques. When one looks at the rapid and intensive inventions and innovations in the field of micro-electronics during the last five years, the conclusion must be that a precise prediction of technological developments during the coming period is very precarious. Nevertheless the organization must present their best view of the state of the art and of the near future to the participants, so that meaningful discussion can be started about the effects of the technologies on the quantity and quality of employment in the organization.

3 Information handling actions: six dimensions

Especially in office work, the essential tasks are all related to information handling. These tasks can be separated into six groups of information handling actions, as Roessner *et al.* (1985) and Porter (1987) have argued.

1 Information intake.

The acquisition of existing information from outside the organization/department by the individual. This also includes screening, reading, listening, and searching outside the worker's information system. Examples of actions are:

- reading
- incoming mail
- information search*
- meet/travel*

2 Information processing.
Reviewing, revising, analyzing, modelling, planning, scheduling, deciding, and selecting. Includes both routine (rule-based) and non-routine information processing (with the former more amenable to automation). Examples of actions are:

- proofing
- calculating
- scheduling

3 Database management.
The storage, retrieval, and update of information within the organization. Examples of actions are:

- filing
- paper handling
- information search*
- typing*

4 Information production and transformation.
Generation of information in the form of memoranda, reports, correspondence, speeches, etc. Also includes the entry of information into a technological system by the worker. Examples of actions are:

- writing
- dictating
- typing*

5 Information dissemination.
'One-way' or 'broadcast' communication (speeches, reports, instructions, letters, etc.). Examples of actions are:

- copying
- telephone*
- meet/travel*

6 Interactive communication.
'Two-way' exchange of information (meetings, conversations, etc.). This is distinct from dissemination by the impact of quick feedback. Examples of actions are:

- telephone*
- meet/travel*

---

* = action in other groups also

4 Time spent in functions

In step three we introduced the six dimensions of information handling function(s). When we use the IDM approach it is necessary to have data on the time that workers spend in the six dimensions. This overview of the spending of time has to be present at the moment that the IDM approach is applied. In many organizations this type of data is not directly available. That means that some research has to be done to collect the data about the time spent for the function(s) that will be analyzed. Mostly time registration studies are the way to obtain this type of data. For each action, such as the examples mentioned in step three, the time spent on that activity during a representative period must be recorded. These actions, with the data about time, are then combined in the six dimensions. Often it is not so easy to decide what action belongs to what dimension, and what part of the time spent must be added to a specific dimension. For example meetings and travelling are activities that mostly belong to more than one dimension. In many cases some inventive decisions have to be made to reach an acceptable level of consensus between the executives and the researchers. Meanwhile there are some research data available that can be seen as a point of reference for the time patterns.

Zanders and Willems (1987) obtained results, summarised in Table 28.2, in the Dutch non-commercial service sector for three occupations: managers, professionals/staff employees, and secretaries.

Table 28.2. Time spent (%) by action dimension for three occupations in The Netherlands (1987)

	Managers	Professionals/staff	Secretaries
information intake:	35	20	27
information processing:	7	17	26
database management:	1	5	10
information production/transformation:	7	16	17
information dissemination:	22	20	15
interactive communication:	28	22	5
Total:	**100%**	**100%**	**100%**

Data for the United States in 1980 demonstrate the pattern as indicated in Table 28.3 (Porter, 1987).

Table 28.3. Time spent (%) by action dimension for three occupations in the United States (1980)

	Managers	Professionals/staff	Secretaries
information intake:	19	21	11
information processing:	18	16	39
database management:	7	9	13
information production/transformation:	10	21	9
information dissemination:	7	9	15
interactive communication:	39	24	13
Total:	**100%**	**100%**	**100%**

5 The EFTE procedure: A Delphi model

Now comes the heart of the matter: the procedure to involve, in a systematic way, employees in the process of estimating the effects of new technologies on their jobs. The approach, developed by Roessner and Porter is named the EFTE procedure, which stands for:

- Estimate
- Feedback
- Talk
- Estimate.

The EFTE approach is a variant of the classical Delphi method. The Delphi approach may be characterized as a method for structuring a group communication process so that the process is effective in allowing a group of individuals, as a whole, to deal with a complex problem (Linstone and Turoff, 1975, p.3). In some views the Delphi approach is fundamentally the art of designing communication structures for human groups involved in attaining some objective (Linstone and Turoff, 1975, p.489). Two versions are possible. The first and the most common is the paper and pencil version which is commonly referred to as a 'Delphi Exercise'. The second and newer form is the 'Delphi Conference', where the monitor-team is, to a large degree, replaced by a computer which has been programmed to carry out the compilation of the group results. The two versions are also labelled as the 'Conventional Delphi' and the 'Real-Time Delphi' (Linstone and Turoff, 1975).

The Delphi method is one of the spin-offs of defence research in the fifties. 'Project Delphi' was the name given to an Air Force-sponsored Rand Corporation study, starting in the early 1950s, concerning the use of expert opinion (Dalkey and Helmer, 1963). The objective of the original study was to obtain the most reliable consensus of opinion of a group of experts by a series of intensive questionnaires interspersed with controlled opinion feedback. The subject in the first Delphi study was the application of expert opinion to the selection, from the point of view of a Soviet strategic planner, of an optimal U.S. industrial target system and to the estimation of the number of A-bombs required to reduce the munitions output by a prescribed amount. More attention was given to the Delphi method in the mid sixties after the publication of the 'Report on a Long-Range Forecasting Study', by T.J. Gordon and Olaf Helmer, published as a Rand paper in 1964 (Gordon and Helmer, 1964; Helmer, 1966). That study was designed to assess the direction of long-range trends, with special emphasis on science and technology, and their probable effects on our society and our world. Long-range was defined as the span of ten to fifty years. Since the mid sixties, Delphi has spread from America to Western Europe, Eastern Europe and the Far East. The rapid change and growth of aerospace and electronics technologies and the large expenditures devoted to research and development leading to new systems in these areas placed a great burden on industry and defence planners. Forecasts were vital to the preparation of plans as well as the allocation of research and development resources, and trend extrapolations were clearly inadequate. As a result, the Delphi technique has become a fundamental tool for those in the area of technological forecasting and is used today in many technologically oriented companies. Even in the field of classical management and operations research there is a growing recognition of the need to incorporate subjective information directly into evaluation models dealing with the more complex problems facing society.

After this short history of the Delphi method comes the EFTE approach. Typical for this approach is that the participants are acting in face-to-face discussions. In the classical Delphi the participants are, and stay, anonymous, but in the EFTE procedure a meeting is organized for all the actors who will give their estimates and prognoses. The procedure is as follows. In a primary round the actors give their first estimates about the effects of technological systems on their functions or jobs. A more detailed explanation about the factors to estimate and the technique that is used is given in the next step, number six.

Then there is an immediate feedback of the results to the participants. That is realized by using a computer package that gives a graphic representation of the first estimates. The participants are operating in a group session, but their opinions stay anonymous, because the presentation of the results is so arranged that the data for the group as a whole are given. The joint graphic representation of the different individual estimates offers the data for discussions among the actors. After these discussions and argument the second round starts with estimates. This process of giving estimates, feedback, talk and again estimating continues until a certain level of stability is reached. That means that there are no more new arguments in the discussions and the estimates show the same pattern in two successive rounds. Consensus is not a necessary condition to stop the process. Different opinions are respected.

6 Three factors to estimate

The estimates are given for three parameters or factors:

A: Time effects
B: Workload
C: Adoption.

Time effects estimates give the changes in working hours that are expected when new information technologies are implemented in organizations. Workload estimates give a prognosis of the expected amount of work in specific jobs in the future. Adoption prognoses give the expected percentage of employees who will adopt the new information technologies in their work. In each EFTE round the estimates are given on the relevant factor for the six dimensions described in step number three.

The sequence of the procedure is that the three parameters or factors are handled in a successive manner. The starting point is the time effects. Different rounds of time estimates are completed until stability is reached. Mostly it appears that three rounds are sufficient. Then the rounds for the second factor, the workload are started. Adoption is the last factor that comes into the picture. At the beginning of the procedure the actors receive a precoded sheet on which they can indicate their estimates. In fact they are asked to mark numbers that correspond closest to their opinion.

In the first rounds questions usually arise about concepts and definitions. These have to be solved during the discussions after the presentation of the first results. So, in the second and further rounds the actors have the same frame of reference for the concepts and definitions.

The descriptions and definitions used in empirical settings are presented here.

A: Time effects

Employees need a certain amount of time to complete information handling actions in their work. New technologies can and will affect that amount of time. The question is what the expectations are about the changes in time when new information technologies are used by the employees. These estimates for time effects must be based on the activities that are executed today. Doing the same work as now, how much less or more time will be needed when new technologies are used to complete that work?

Often the participants include here the changes in workload that are related to the introduction of new technologies, but these effects must not be included in the estimates about time effects; these are a part of the workload estimates. In most cases, however, this analytical distinction is hard to handle in reality.

In the original EFTE design the Georgia Tech. researchers used the concept of time savings instead of the more neutral term time effects, but the concept of time effects is preferable because the suggestion that new technologies always result in a saving of time is not correct. Especially during the first period when new technologies are used sometimes more time is needed to complete the same amount of work. The savings may often come later, after a learning period.

B: Workload

Workload refers to the amount of work that has to be done in a function. The question is whether changes in the amount of work are expected in the near future and to what extent. These changes in the amount of work, or workload, can be affected by a broad range of factors; technological, economic, organizational, political etc. Here it is not only the technology that has to be involved in the process of estimating changes; all the factors are relevant that can have an influence on the changes in the amount of work.

C: Adoption

Technological systems can have an effect on the time patterns of workers. However, that holds good only when the new technological systems that are available for and in an organization will be accepted and used by the employees, so estimates have to be made about the percentage of workers who will use the new technologies. Here then the technologies must be considered as applied to estimating the time effects. The question is how many employees will work with the new technologies.

7 Execution of the EFTE procedure

Before we apply the EFTE procedure in an organization it is necessary to take some decisions about practical affairs. First an explicit choice must be made about the forecasting period, both the base year and the projection year(s).

The base year is usually the actual year in which the procedure is applied. The projection year(s) is a point for discussion. First a decision has to be made if one or more projection years are wanted. It has been mentioned already that the introduction of new technologies is often coupled with a learning period, so that it takes some time before the advantages of the new technologies can be fully utilized. Therefore it can be informative to select two projection years; e.g. years from the base year and then four or five years from the base year. If one projection year is wanted it is advisable that projections are made over a period of three to five years. Looking much further into the future makes the estimates less realistic.

A second choice that has to be made is about the character of the estimates. Three options are open: minimal estimates, most likely estimates, maximum estimates.

In many cases it is sufficient to ask only for the most likely estimates, but in future studies it is often usual to present also minimum and maximum trends in the prognoses. That can also be done in the EFTE procedure, but it is optional.

8 Calculation of the total effects

The IDM approach ends with the presentation of the total effects, based on the estimates for the three factors: time effects, workload and adoption. The combined effects are calculated with the formula:

$$WORK_t = WORK_b \times W_t \times ((1 - T_t) \times A_t + (1 - A_t))$$

This formula can be reduced to:

$$WORK_t = WORK_b \times W_t \times (1 - (A_t \times T_t))$$

Whereby:

$WORK_t$ = Employment at the future time *t*.
$WORK_b$ = Employment in the base year *b*.
W_t = Workload at time *t*.
A_t = Adoption of the new technology at time *t*.
T_t = Time saving effects at time *t*.

Use of the formula indicates to what extent the work in the near future will be affected by new technologies according to the estimates of the participants. The session ends with figures that give the separate estimated effects on time, on workload and on adoption and with a figure that gives the combined total effect.
Normally a session with fifteen to twenty participants takes about six to eight hours depending on the number of discussion rounds that are needed. The session ends, as already mentioned, with figures, but the main aim of the IDM approach is not to present exact figures, but to involve employees in a process of thinking about the opportunities and the threats that new technologies have to offer to information workers.

Conclusions

During several practical sessions, experience has been gained with respect to the application of the IDM approach, on the basis of which the following conclusions can be drawn:

- The IDM approach is in the first place an awareness and implementation method.
- The method stimulates the direct involvement and participation of employees in innovations with new information technologies.
- The method offers a systematic strategy to analyze the expected effects of automation on employment.
- Data become available that are comparable within and between organizations.
- Results are immediately available, while all the input is directly processed in the IDM computer package.
- The IDM approach offers data for a socio-technical plan that can be integrated in an information and corporate plan.

A last point is the following. Until now the IDM approach has been directly linked to automation and information systems. However, this interactive Delphi approach can also be applied in a more general way to different subjects. Nelms and Porter (1985) mention the wider perspectives of the Interactive Delphi Methodology. Exploring these wider perspectives, for example in the field of human resources planning and future job profiles, is now in process. It is hoped and expected that the IDM strategy will be explored and be used by many people and organizations. The proof of the pudding is in the eating.

References

Buyse, J.J. (1987). *Quality of work and organization on the shop-floor. An integrated study on micro level*(University of Nijmegen) (In Dutch: Kwaliteit van werk en organisatie op de produktievloer. Een integrale studie op mikro-nivo)

Dalkey, N. and Helmer, O. (1963). An Experimental application of the Delphi Method to the Use of Experts. *Management Science* **9**, No.3

Daniëls, J.J.M.C. and Duijzer, G. (eds.) (1988). *Delphi: Method or mode*? (SISWO: Amsterdam) (In Dutch: Delphi: Methode of mode?)

Eijnatten, van Frans, M. (1985). *STTA: Towards a new work design paradigm* (University of Nijmegen) (In Dutch: STTA: Naar een nieuw werkstruktureringsparadigma)

Gordon, T.J. and Helmer, O. (1964). Report on a Long-Range Forecasting Study. *Rand Paper* P-2982

Helmer, O. (1966). *Social Technology* (New York: Basic Books)

Helmer, O. and Rescher, N. (1960). On the Epistemology of the Inexact Sciences. *Project Rand Report* R-353

Houten van, H.J. (1988). Five variations on the Delphi theme. In *Delphi: Method or mode?*, edited by J. Daniëls and G. Duijzer (Amsterdam: SISWO), pp 5–35 (In Dutch: Vijf variaties op het Delphi-thema)

Hunt, H.A. and Hunt, T.L. (1986). *Clerical Employment and technological Change* (Kalamazoo, Michigan: W.E. Upjohn Institute for Employment Research)

Linstone, H.A. and Turoff, M. (eds.) (1975). *The Delphi Method. Techniques and Applications* (Reading, Massachusetts: Addison-Wesley Publishing Company)

Mumford, E. and Weir, M. (1979). *Computer Systems in Work Design: The ETHICS Method, Effective Technical and Human Implementation of Computer Systems*. A work design exercise book for individuals and groups (Associated Business Press)

Nelms, K.R. and Porter, A.L. (1985). EFTE: An Interactive Delphi Method. In *Technological Forecasting and Social Change* **28**, 43–61

Porter, A.L. (1987). A Two-Factor Model of the Effects of Office Automation on Employment. *Office Technology and People* **3**, 1, 57–76

Porter, A.L., Roessner, J.D., Rossini, F.A. and Nelms, K.R. (1985). *Office Automation Outlook: 1985-2000*. Report to the Office of Technology Assessment, U.S. Congress (Atlanta: Georgia Institute of Technology)

Roessner, J.D., Mason, R.M., Porter, A.A., Rossini, F.A., Schwartz, A.P., Nelms, K.R. (1985). *The Impact of Office Automation on Clerical Employment, 1985-2000*. Forecasting Techniques and Plausible Futures in Banking and Insurance (Westport, Connecticut/London, England: Quorum Books)

Roessner, J.D., Mason, R.M., Porter, A.L., Rossini, F.A., Schwartz, A.P., Sassone, R.G., Tarpley, F.A., Schaetzel, T.N., Nelms, K.R. and Diehl, S.G. (1984). *Impact of Office Automation on Office Workers. Volume III*: Technical Report. *Volume IV*: Appendices. Report to the U.S. Department of Labor (Atlanta: Georgia Institute of Technology)

Sitter, de L.U. (1982). *On the way to new factories and offices* (Deventer: Kluwer) (In Dutch: Op weg naar nieuwe fabrieken en kantoren)

Sitter, de L.U. *et al.* (1986). *The flexible company* (Deventer: Kluwer) (In Dutch: Het flexibele bedrijf)

Zanders, H.L.G. and Willems, A.G. (1987). *Automation in the Non-Profit Service Sector, Consequences for Employment* (The Hague: Office of Education and Sciences)

Part VIII

New technology issues

Although many of the earlier contributions draw upon work in, or make recommendations about, technological change, this last section is dedicated to contributions specifically on new technology issues. The importance of these for ergonomists cannot be overstated. As companies slowly realise that the return on investment, indeed the very viability, of their technological innovations depends upon how they manage their human resources and on their human factors design generally, then so ergonomists will be expected to come up with 'solutions.' We will be expected to be able to provide critical input for new production control systems, group working processes, training and job aiding, knowledge based systems, interfaces and systems implementation.

Our final set of contributions starts with just such a comprehensive view of human factors needs with new systems. Bradley describes a Swedish programme, RAM, and a 'tree of action strategies' for company, national and international strategies as regards human issues in the move to what she terms knowledge based systems (KBS). Her particular concerns are the psycho-social objectives to be achieved during computerisation, and the problems that can occur with the organisation structure, work roles and job content, knowledge needs and decision making, and skills, education and training.

Martin in the next chapter also advances a framework within which to study psycho-social effects, in what is called advanced manufacturing technology (AMT). His framework focuses on four job dimensions—control, cognitive demand, cost responsibility and social interaction. Of great interest in this is how these four dimensions might be expected to relate to key outcome factors of system performance, operator stress and job satisfaction. Martin offers at least initial support for his framework by reporting and interpreting results from a recent study of CNC machine operators.

The next two contributions take slightly sideways looks at new technology, but raise some vital issues in doing so. Shotton's study deflates some of the myths which have grown up around so-called 'computer junkies.' From this in-depth examination of people (almost all males!) who spend long periods computing, Shotton draws lessons for the design of computer equipment, expected health effects and good job design. Craven in the chapter which follows also expands the focus of our concern. He emphasises the need to account for different cultures, and consequently expectations, knowledge and social systems, when designing or introducing new technical systems in countries other than one's own. Training, in all its forms, he regards as a vital pre-requisite to effective new technology implementation.

Finally, Fruytier and ten Have discuss organisational influences in technology change. They compare cases from two companies changing to CNC-based production. The most noticeable difference between them is that operators in one company have much greater freedom to adapt programs and control the process, and have more non-CNC tasks involving determination of quality actions, training new workers, and assisting in performance measurement. Results of the studies were that the better consideration of operator roles in the first company had led to improvements in flexible development of personnel and quality of production. Moreover, the investigators argue that this company has a considerably improved organisational and work systems structure.

If anything should be drawn out of these studies it is that the successful integration of new technology and new jobs will depend upon an individual company's philosophy and structure and consequent human resource policies. The responsibility for ensuring that companies show adequate concern for human factors in their policies, and that these policies are implemented successfully, lies with ergonomists and other work redesign professionals.

Chapter 29
Knowledge based systems and work design

Gunilla Bradley

Institute of International Education, University of Stockholm, S-106 91 Stockholm, Sweden

Introduction

This chapter describes a new research programme at Stockholm University. Its aim is to study the introduction of knowledge based systems (KBS); consequent changes in the organizational and psychosocial work environment are being described and analyzed. These changes will have a deep impact on society as KBS come into widespread use. In addition to the author's professional background in educational psychology and applied sociology and her research in computerized work and women's work, the research programme is underpinned by Swedish traditions in working life and the labour market. These are now manifested through laws and agreements on quality of work life and co-determination, based on a collaborative atmosphere between labour and management. Also, an international comparative approach was taken.

It is the view of the author that the ultimate aim of research in this field should be to improve the quality of life for everybody. Therefore this work:

- lists psychosocial aims during computerization in general, derived from findings of the RAM project (see below)
- presents several strategic problems during transition to KBS
- presents a sample of hypotheses for future research

RAM programme

Different data-processing systems offer different conditions for the structure and design of work and its organization. Three main types of data-processing systems were studied by Bradley during the RAM programme (at the University of Stockholm, 1974–1986). These three systems are related to three phases in the history of computer technology:

1. A batch-production system at a state-owned company
2. An on-line system with display terminals at an insurance company
3. A microcomputer system at an electronics company

Problem areas concerned the computerization of working life. Essential concepts derived from a theoretical model were used to study the following: general questions about the work environment and work satisfaction; information and participation in decision-making; organizational design, work content and work load; promotional and development patterns;

contact patterns and communication; salary conditions and working hours; education and training; evaluation of work roles; physio-ergonomic conditions; leisure time and health. Theories, methods and results from the RAM programme are summarized in the book *Computers and The Psychosocial Work Environment* (Bradley, 1989).

Psychosocial aims during computerization

A number of aims have been derived from the RAM project which are relevant to research on KBS. The basic object is to achieve a work situation and work content that fulfil the needs and requirements of each and every individual at work. In more detail we aim to have:

- Neither too little nor too much responsibility, variety, attention and complexity in the work.
- Working pace adjusted to the individual and not too dependent on the pace of workmates and/or equipment.
- Possibility of influencing and changing the organization and planning of tasks, working hours, breaks, variation in pace of work, etc.
- Good flow of information among staff, both vertically and horizontally, e.g. instructions, company policy, etc.
- Special support for staff who have a 'buffer role', avoiding the negative effects on staff in this position if they are faced with contradictory demands.
- Development of personnel on the job: emphasizing continuous personal and professional development for the individual.
- Adequate training during paid working hours.
- Possibility of conferring with colleagues in order to fulfil needs for human contact and communication during daily tasks.
- Minimum of physical strain on eyes, neck, back, muscles, etc.
- Opportunities to replace workers on the job; it should not be too easy or too difficult to replace them, however. Positions entailing too much specialization may need to be avoided.
- Extrinsic equality between the sexes; an even distribution of duties and positions.
- Intrinsic equality between the sexes: planning new job functions so as to utilize both traditional female and male skills, knowledge, interests and values.
- Enlightened and steadily improving rewards and promotion systems at every level, including management.

Problems in the change to knowledge based systems

The consequence of companies changing to the use of KBS is that a number of transitional concepts and operations must be re-examined and re-defined. Many problems will arise and questions will crop up as outlined below. These were formulated during the RAM project, from a study of CAD and engineers' roles (Bradley, 1987), in the US pilot study discussed below, and through the general KBS literature.

Organizational structure

1 What will be the most obvious changes in the formal organizational structure of a

company resulting from the introduction and use of KBS? What new concepts will be needed?

2 How can we then achieve the desired organizational structure (based on scientific research and Swedish labour legislation), at the same time taking into account the opportunities and limitations that accompany the use of KBS?

The content and organization of work roles

1 How can we describe the work content in new occupations that arise as KBS are introduced? E.g. knowledge engineers and leaders of KBS projects.
2 What changes in work content can be observed for existing occupations involved in a KBS project? E.g. experts and users (individuals, work groups, managers).
3 How will occupational roles develop in relation to KBS usage? What trends are desirable?

Job content/the nature of work tasks

What changes take place in job content during the use of KBS? Those aspects considered to be the most crucial are:

- the opportunity to use one's own knowledge and ideas
- the possibility of following a job from beginning to end, and of seeing how one's own work forms part of the company's overall activities
- the level of complexity/degree of difficulty required in the work
- the level of professional qualifications required by personnel
- the degree to which an individual can be replaced
- the opportunity for workers to influence methods of working, design of the work, and planning.

Knowledge requirements—education, influence, and decision-making

(1) What are the knowledge or educational needs of those involved in a specific KBS (experts, knowledge engineers, project leaders, users)?

  a What type of training should be given within the company (e.g. in teamwork, problem-solving, decision-making)?
  b What type of education should be given by society?
  c What type of motivation is needed in a society where KBS are widely used?
  d Which pedagogical principles should be applied?

2 How can we describe the dominant *pattern* within a company as to *knowledge*: its transfer, distribution and growth-knowledge tied to both people and computers?
3 In parallel terms (authority, power, etc) how can the transfer and distribution of personal influence within the company be analyzed?
4 How can we develop a map or method for analyzing and documenting where decisions are made in a company, including decisions made by people and decisions built into the computer?

Sample of hypotheses

During a period as a guest researcher at Stanford University, USA, in 1987, the author conducted a pilot study. This comprised semi-structured interviews with researchers within computer science and the behavioural sciences, and with people employed at high tech. companies, people with early experience of the use of KBS in working life. These persons were confronted with the questions and problems formulated within the RAM research.

A number of assertions and hypotheses were derived from these interviews; these are now being used as points of departure for empirical research in the current Swedish programme.

Organizational structure and design

- A company that learns how to develop its own KBS will experience a deep and fast impact on its own structure. However, this will require a 'success story' at the beginning of the introduction process. (There will be a ripple effect.)
- The hierarchical structure of a company will diminish with the introduction of KBS. When computerization takes place both from the bottom upwards and from the top downwards in the company, the middle level will grow.
- 'Ambiguity' in organizations will be more pronounced, e.g. ambiguity related to the limits of 'competence' of the machine and the knowledge of human beings.
- The concepts of 'power', 'influence', 'authority', and 'participation' in the work place must be re-defined as understanding increases about the psychosocial aspects of using KBS.
- A company that has a formal and rigid definition of work roles will have a slower rate of introducing a KBS. An 'unsuccessful story' is more likely to result than in a company where work roles are more flexible and less formal.
- To be able to cope with the KBS phase of computerization there will be more need of cooperative patterns at the work place, thorough review of salary principles and policies, non-hierarchical structures, use of interdisciplinary and interdepartmental contacts.
- Using the terminology of the psychology of work behaviour, the challenge of this technology is to permit the possibility of workers seeing the impact of their own work effort on the total process of production.
- Combining KBS with traditional computers and telematics will strengthen the organizational structure that is characterized by networking, flexibility and non-hierarchy.

Work content

- If the KBS is very successful, there will be fewer skilled people needed to do the work. A small superskilled élite will result (automation from the top). In ten years' time, however, it will be difficult to get a human expert (a person instead of a system) in certain fields to collaborate in the development of new KBS.
- The introduction and use of KBS have not been perceived by the persons involved as entailing a large general change. Rather, they are perceived as a new tool and people are not worried by them. There is an 'invisible revolution' which often concerns 'islands' in the company, with low collective identity. The invisibility is also

connected to the fact that more extensive usage of KBS occurs during that phase of computerization often named micro-computerization or the use of distributed computer power.

- There are two tendencies as regards the competence of the experts:
  - Experts will be 'liberated to do the more difficult tasks', thus living more interesting lives.
  - Experts will work harder to learn new skills.

  These two tendencies are related to the expert's age, role within the system and point within the system life cycle, and degree of emancipation as to sex-role patterns.
- There will also be two tendencies, alternative this time, regarding the competence of workers (non-experts) using KBS:
  - Workers will be more valuable to the company and gain more competence because they are taught by the KBS.
  - Workers will be less valuable to the company and lose relative competence, since the company will train more people on the system to do a certain job.

Skills - education and training

Skills in collaboration

When a company introduces KBS, participation (influence) involves all categories of professions and persons, e.g. the end user, the domain expert, the knowledge engineer, and the project manager.

Six types of educational requirement will be very important throughout the company:

- Learning to work in teams (training at school is individual, while most work in the future will be teamwork)
- Learning to solve problems
- Learning to plan
- Learning how to make decisions
- Learning about women's and men's jobs
- Learning to work 'between cultures'.

Transfer of skills

- Skills in a company can now rise (be driven upwards) and be concentrated upon the experts themselves ('Expert use of expert systems')
- Skills in a company can now descend (be driven downwards) ('Worker use of expert systems')

Successful use of KBS requires a plan for educating and training the work force, a plan for job security, and for alternative careers within the professions involved. This is valid for experts, end users, internal knowledge engineers and other workers. Quick supportive training of an innovative kind is needed for workers. More extensive use of KBS requires more generalists. Specialization will continue but interdisciplinary skills must be developed. Versatile workers with a good overview will be needed to create new syntheses.

Conclusions from US study and RAM programme

The cognitive issues within KBS research have so far received most attention. Motivational issues, on the other hand, have been little studied, but they are especially important. The experience from the pilot study in the USA suggests that the successful implementation of KBS depends to a high degree upon motivational and psychosocial issues.

New aspects of the psychosocial work environment are emerging. Both the number of issues and their multi-dimensionality will increase. Computerization at the lower levels of a company has changed the psychosocial work environment. However, the use of KBS also affects the psychosocial work environment of the well-educated, where psychosocial problems dominate over the physical ones.

A KBS influences the transfer of knowledge within a company. Both directly and indirectly, it thus shapes the distribution of power and influence. Some KBS are called decision support systems. It is of special importance both to develop KBS software and to study the introduction and use of KBS in a Nordic country like Sweden. Swedish co-determination law and the associated agreements regulate decision-making in work life. The advanced experiences of working democracy and economic democracy in Yugoslavia may also be very important within this research field.

Discussion of actions

Action strategies—an overview

Figure 29.1 depicts possible relevant action strategies. The 'tree' is based on research by many people in Sweden, and shows actions at various levels: national level (trunk), company level (branch), and individual level (branch). Technology is yet another branch. All the actions are interacting or at least ought to interact, according to the RAM programme's theoretical model for the work environment.

This picture was presented at two previous conferences. On both occasions the audience responded with lively comments—maybe because it was simple yet described a complex phenomenon. All readers of this book are invited to forward comments and questions about this tree, its branches and sub-branches to the author. In particular, answers are needed to the following questions:

What are the experiences in other countries?
What actions at these levels are going on in various countries?
What are the main interplay and interactions?
What are the main controversies and contradictions?
Are there any branches or sub-branches that we have not yet identified?
Where do we need more research?
Which is the most urgent matter for international collaboration?
How and at what level and/or branch must we collaborate?

Can we all work together and supplement our national trees by an international tree of action? Such a tree would show what is going on in many countries. We can then identify the strategies of international actions. Our aim should be to strive for an information society that could be depicted as a tree of fruits for us all to share.

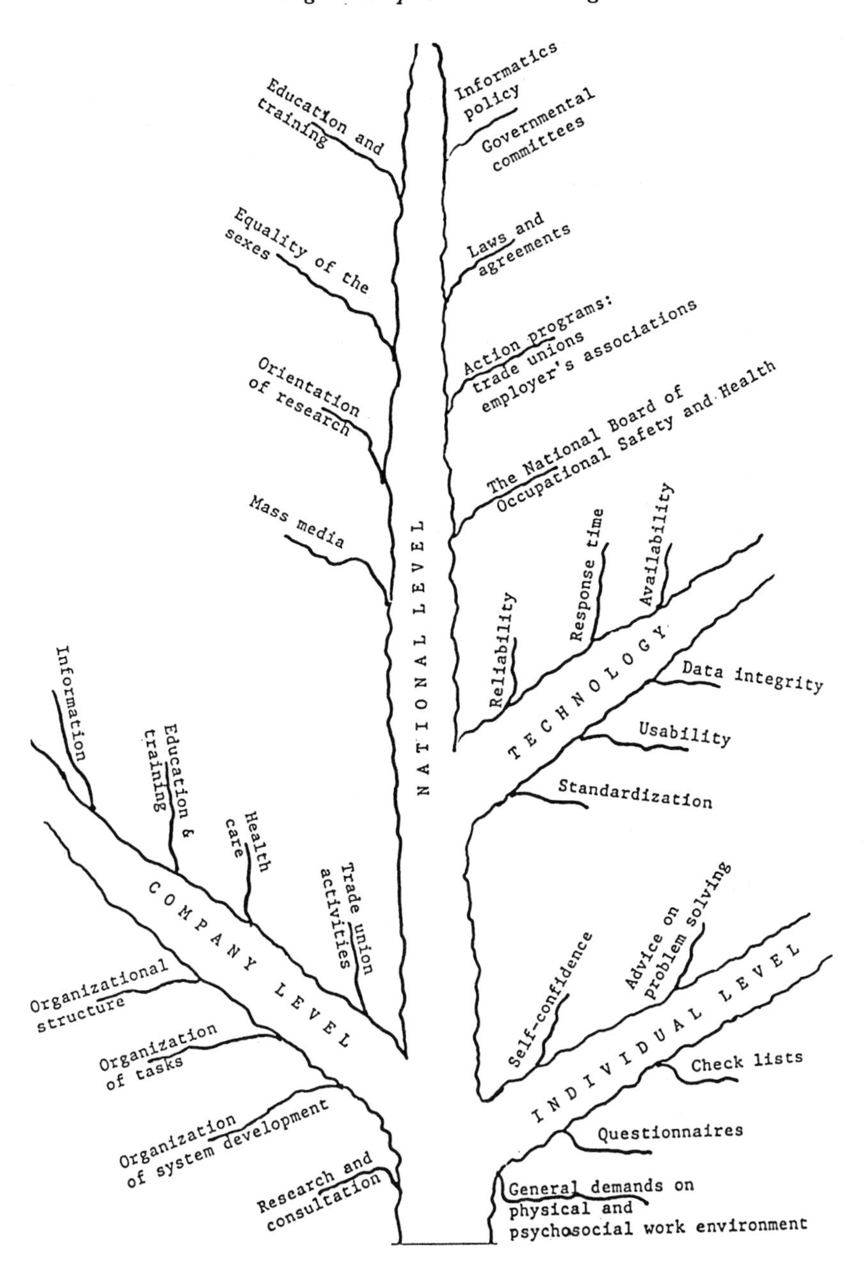

Figure 29.1. Action strategies in Sweden

Some actions at the national level

During 1980–1985 Sweden had a Data Policy Commission and for some years, a 'Minister of Information Technology'. The present government has followed up this work and presented a Government Bill on Informatics Policy in the summer of 1985 and in 1988 the Swedish Information Policy was made public.

One of the most complex labour market issues in Sweden during the 1970s was employee participation in decision-making at the work place. A long series of changes and innovations took place. Three different levels of participation are usually mentioned: shop-floor participation, company participation and financial participation. (Corresponding index measures were used in the RAM studies.) Shop-floor participation refers to the ultimate purpose of co-determination reform, e.g. to enhance the individual's influence over his or her own working situation. Working systems are one area in which companies and government agencies have shown great interest. Swedish companies like Volvo and Saab-Scania have attracted attention because of their efforts to improve the design of work places and job content, so as to create a better working environment and increase the opportunities for employees to influence their work situation. Company participation refers to the influence of employees over their work place through representative channels. In Sweden, this is done entirely through the trade unions. Employee representation on boards of directors was introduced in 1973. Its aim was to give the unions increased information while giving boards of directors the benefit of experience accumulated by employees.

A paragraph of the Co-Determination Act regulates the employer's primary duty to keep the employees informed about planned actions. Another paragraph defines the employer's duty to negotiate with the trade unions. These paragraphs should be given particular consideration in the context of computerization. The duty to negotiate spans a broad field, including various forms of computer use and problems connected to systems development. The Co-Determination Act has been supplemented by separate agreements.

The Working Environment Act was introduced in 1978. The concept of the working environment now encompasses work systems, work hours, and adaptation of work to human factors, both physical and psycho-social. Safety stewards (delegates) are guaranteed the right to halt dangerous work under certain circumstances.

The National Board of Occupational Safety and Health has given top priority to certain matters, one being computerization in relation to working environments. Knowledge gained from research and general experience constitutes the basis of new laws and regulations, developed in cooperation with all the different interests involved in collective bargaining between employers and employees.

In the 1985 Government Bill on Informatics Policy, educational questions were given priority—at both basic and advanced levels. The Swedish Government allocated funds in 1984 to be used to attract more women into industry and the technical professions. The campaign was one step in a long-term plan to broaden the labour market for women. It is being conducted in project form throughout the country, using three approaches:

- Measures for influencing girls' educational and professional choices
- Computer education/training for women
- Recruitment and continued educational measures.

Actions at the organizational and individual levels

A major part of the current international debate on computerization deals with the struggle for influence and power over the development of computer applications. Scientific reports and popular books analyze different strategies for system development. Theories have been presented with the aim of replacing, through a democratic process, the one-sided nature of planning and execution of rationalization which has dominated until now.

A parallel process involves identifying, structuring, describing and analyzing work environment problems related to organizational and psycho-social matters. Awareness and recognition of work environment problems and their underlying causes are increasing. Desires and demands, expressed in actual programmes, are receiving wider attention.

People at work can present their preferences and requirements for the work environment when workplaces are computerized. General requirements may be laid down for an acceptable work environment, along with group-based requirements for a particular industry, a particular trade union, etc. Checklists may be used to elicit suggestions on which environmental issues should be considered, during the process of computerization. Questionnaires or interviews may be used to evaluate planned computer systems or systems already in operation.

When data-processing is introduced at their workplaces, workers must express their wishes, as individuals, for future development. They must be able to answer questions such as:

- What conditions of life (leisure time, family life, social life outside what is now called work) do they wish to have in the future?
- What are the desirable characteristics of a good work environment?
- What is the desirable content of their future professional roles?

There is an interaction between different levels of the work environment. Also, at a personal level the individual is affected by the environment, but the individual too has an effect upon his or her environment. A number of instruments have been established at the national and corporate levels. Conditions have been created for broad action to improve the work environment and work itself to bring about corporate change, but it is up to workers themselves to transform their work and its content, both in terms of desirable goals and ways in which they can work towards attaining these goals. Thus this interplay is also an important factor when action is taken in this field. Just as important as the existence of laws and agreements is what people themselves as individuals attempt to achieve on the basis of these laws and agreements.

General demands on the organizational and psycho-social work environment were derived from the RAM project; they were summarized at the start of the paper and cover working pace, influence, development on the job, education/training, human contact and communication, information, intermediate position/buffer role, physical strain, replaceability/job security, extrinsic and intrinsic sex equality. Computerization should also help to create work environments where as few individuals as possible are overstimulated or understimulated. It is the individual who ought to make this evaluation. There are both quantitative and qualitative aspects of overstimulation and understimulation. The number of tasks to be done per unit of time concerns quantity. The degree of complexity is a qualitative aspect.

Referring to the overview of the actions on new technology and work life, the most challenging task is to find a balance between strategies at the different levels or 'branches in a tree of action', not least between efforts at the national level and individual level.

References

Bradley, G. (1987). Changing Roles in an Electronic Industry—Engineers Using CAD System and Secretaries using Word-processing System. In: *Social, Ergonomic and Stress Aspects of Work with Computers*, edited by G. Salvendy, S.L. Sauter and J.J. Hurrell Jr. (Amsterdam: Elsevier Science Publishers B.V.) pp. 295–302.

Bradley, G. (1989). *Computers and the Psychosocial Work Environment* (London: Taylor & Francis)

Chapter 30
A framework for studying the psychological effects of advanced manufacturing technology

Robin Martin*

MRC/ESRC Social and Applied Psychology Unit, University of Sheffield, Sheffield, S10 2TN, U.K.

Introduction

Advanced Manufacturing Technology (AMT) refers to computer-based technology, ranging from 'stand alone' computer numerically controlled (CNC) machine tools and robots to large scale flexible manufacturing systems (FMS) and computer-integrated manufacturing (CIM). For the majority of organizations the smaller-scale 'stand alone' applications are the reality, and are the focus here.

Recently, an emerging literature has developed concerning the job content implications of AMT. Much of this has been driven by the de-skilling thesis (e.g. Braverman, 1974) which predicts that AMT results in a reduction in the skill content of shopfloor operator jobs. Whilst in practice this often does occur (see Butera and Thurman, 1983) this outcome is neither inevitable nor so deterministic. It is clear that there are variations, and thus choice, as to what is the content of AMT jobs (Clegg *et al.*, 1984).

The literature on AMT has raised a number of important issues but its emphasis on skill, whilst no doubt an important factor, has so focused enquiry that other potentially important properties of AMT have been ignored (Wall and Martin, 1987). The current framework has been developed to address these issues (Wall *et al.*, 1990a). Drawing from the job design, stress and AMT literatures four main job dimensions of salience to AMT operator jobs have been identified; control, cognitive demand, production responsibility and social interaction. The framework further describes the relationship between these job dimensions and three main outcome factors namely, system performance, psychological strain and job satisfaction. The framework is described in the following section. Following this, some propositions arising from the framework are tested in some recent research.

AMT theoretical framework

AMT job dimensions

The framework focuses on four main job dimensions which incorporate more specific job characteristics. The dimensions were included in the framework if they satisfied three main criteria. First, they should vary across applications of AMT. Secondly, the characteristics

---

Now at: International Centre for Child Studies, Ashley Down House, 16 Cotham Park, Bristol BS6 6BU, U.K.

should be open to change. Finally, they should be of known psychological importance. These dimensions are introduced briefly below.

Control

The most commonly identified job content implication of AMT concerns the degree to which the operator has control over his or her work. Such a consideration underlies the work on the de-skilling effects of AMT discussed above. The use of control, in this respect, however, has typically been ill-defined. The framework thus identifies three aspects of control which can vary markedly between applications.

Timing control—This refers to the degree by which the operator can control *when* to complete required tasks. More specifically, it covers the extent to which the operator can decide when to carry out the tasks, rather than merely responding to technically determined demands.

Method control—This aspect of control reflects the amount of choice an operator has in *how* to complete tasks. It refers to the extent to which an operator can determine how to complete work, rather than working to set procedures.

Boundary control—The final dimension of control is the extent to which the operator is required to conduct the secondary activities (such as machine maintenance and quality inspection) in support of the primary tasks. It refers more to the role of the operator with respect to managing work.

Cognitive demand

It has been widely accepted that AMT tends to reduce the amount of physical activity required in a job, whilst increasing the cognitive demand upon the operator (Child, 1984). This demand tends to fall into two clear categories.

Monitoring demand—A common characteristic of AMT is that the operator is required to spend long periods of time monitoring the production process. In many situations the operator is a 'machine minder' whose main role is to stop the machine when errors occur. The level of monitoring required varies according to such factors as the design of the technology and its reliability.

Problem-solving demand—Another aspect of cognitive demand is the extent to which the operator is required to solve problems which arise. The extent to which this occurs is a function of the nature of the production system and the distribution of problem-solving skills in the organization. In many situations an operator is required to call specialists (e.g. maintenance, quality control) when problems occur, whilst in other situations he or she is trained to be more active in solving problems (Clegg *et al*., 1984).

Cost responsibility

In many cases, AMT significantly increases the responsibility the operator has for both valuable machinery and products. An error by the operator can have serious consequences. Three (interrelated) consequences of an operator error can be identified. First, an error can lead to damage to the machinery. Typically, AMT is very complex and because it is often tailor-made to the organization's needs, is very expensive. Second, an error leads to lost product. In some situations, AMT systems can increase production up to 40 times. The cost of an error in terms of lost materials is, therefore, correspondingly increased. Third, the correction of errors leads to lost production time. This loss is greatly magnified by the production potential which these machines offer. Combined, these factors lead to situations where operators experience considerable responsibility.

Social interaction

The final characteristic to which attention is drawn concerns social interaction. Of the characteristics described so far, social interaction is the one which has received the least attention in the AMT literature. Two aspects to social interaction are identified which are particularly affected by AMT.

Social contact—AMT systems may affect the availability of social contact in several ways. Most immediately, they may radically affect the layout of the shopfloor, such that the physical distance between employees is increased. In another aspect, AMT may involve a high level of visual monitoring which ties an operator to the machine and thus restricts freedom to leave it. Finally, the increased reliance upon information-processing mechanisms can lead to a reduction in the need for face-to-face communication.

Social support—AMT systems can also affect the quality of social interactions. Operators often become isolated, thus reducing the degree of social support available. Furthermore, there is often less need for reliance upon co-workers and increased need to communicate with specialists such as engineers and programmers.

Relationship between AMT job variables and outcomes

In this section are briefly specified some of the relationships between the four job dimensions described above and three key outcome factors, namely system performance, operator stress and job satisfaction.

AMT job dimensions and system performance

An important index of system performance in AMT is machine utilization (that is, the amount of time the machine is available for production). Typically, the speed of production is set by the technology and therefore output is a function of the amount of time the machine is working. Machine utilization is affected, amongst other things, by the frequency of errors which result in stopping production and by the time taken to correct them.

The predictions here draw upon socio-technical theory's concept of 'control of variance at source'. This predicts that variances (that is, unplanned deviations) should be controlled as close as possible to their source of occurrence. Obviously, the more quickly errors are corrected, the greater will be the resulting machine utilization.

The people who are in the best position to respond quickly to errors are the operators. They are able to make corrections immediately without having to wait for specialists (such as engineers and programmers) to arrive. The framework predicts that the more the operator has control (particularly greater boundary control) for correcting errors, the greater will be the resulting system utilization. However, these benefits are likely to be more effective in AMT systems which require many error corrections (as there is simply greater scope for improvement) than those systems which require relatively few.

AMT job dimensions and psychological strain

With reference to the wider literature in Occupational Psychology, all the dimensions described in the framework have been implicated as causes of psychological strain when in their extreme forms (Warr, 1987). Jobs characterised by extremely low levels of control, excessively high cognitive demand and cost responsibility, as well as lack of opportunity for social contact are reported as stressful. Extreme forms of these characteristics, however, are not typical since their effects are usually so detrimental that such job designs are avoided.

Therefore, for the majority of jobs, which are moderate in these dimensions, a different causal model is required. With respect to this, a number of researchers have adopted a synergistic approach as an explanation of psychological strain, that is, strain results from co-occurrence of specific factors (Broadbent, 1986; Clegg *et al.*, 1987; Karasek, 1979). The principle underlying this causal model is that combinations of job dimensions produce strain greater (or less) than their separate parts. This is sometimes referred to as an 'ordinal interaction'.

AMT job dimensions and job satisfaction

With respect to job satisfaction, predictions of the framework are similar to those found in the wider literature. Research has consistently supported a main effects explanation of job satisfaction and it thus seems appropriate that such a model should apply to AMT jobs. Therefore job satisfaction should increase as an additive function of greater control (timing, method and boundary) and problem-solving demand, lower monitoring demand, and higher cost responsibility, social contact and support.

Some tests of the propositions

In this section we briefly describe some of our recent studies which test three specific predictions arising from the framework.

Operator control and system performance

The study focuses upon 19 operators of CNC automatic insertion machines used to insert electrical components into printed circuit boards (PCBs) (Wall *et al.*, 1990b; Martin and Jackson, 1988). The PCB is secured to a base which is moved into the correct location for the insertion of each component. Components are 'fed' to the insertion head from bandoliers. The complete sequence is under computer control.

Machine breakdown data revealed utilization (and hence performance) was dependent upon the speed with which machine errors were corrected. In this respect the operators' role was relatively limited. When errors occurred, they simply called out engineers, who, because of their heavy work schedule, were often delayed in attending to the machine.

Drawing from the framework described earlier, the jobs were redesigned so as to give the operators greater (boundary) control over their work. More specifically, they were trained to carry out a number of error correction tasks which hitherto had been the responsibility of maintenance personnel. The effects of these changes on performance were recorded for a period of fifty days before and after the job changes.

The results of the job changes confirm the predictions of the framework. Greater operator control over error correction led to a 28% reduction in lost production time and a 40% reduction in the frequency of system breakdowns.

Sources of psychological strain

The framework predicts that psychological strain results from the co-occurrence of certain job dimensions. For example, strain would result from jobs which require a high level of attention (monitoring) as well as high cost responsibility. High levels of either factor alone should not be implicated as a source of strain.

This prediction was tested in a cross-sectional survey of 128 operators in a variety of computer-controlled technologies (Martin and Wall, 1989). The operators were responsible for making the 'bare boards' into which electrical components were inserted.

On the basis of measures of attention demand and cost responsibility, the operators were divided into four groups; low on both factors, high on one and low on the other and high on both.

The results showed that jobs which were high in either attention demand or cost responsibility had no demonstrable effect upon reported operator well-being. However, as predicted by the framework, jobs which were high in *both* attention demand and cost responsibility were associated with greater pressure, job-related anxiety and worse mental health.

Operator control and job satisfaction

The framework predicts that job satisfaction increases as a function of operator control. This was tested in the field experiment described above concerning system performance. Operators' attitudes towards their work were measured before and after their jobs had been redesigned to increase operator control.

The results supported the framework. Greater operator control resulted in improved perceptions of control over their work and greater intrinsic job satisfaction.

References

Braverman, H. (1974). *Labor and Monopoly Capital.* (New York: Monthly Review Press).

Broadbent, D.E. (1986). The clinical impact of job design. *British Journal of Clinical Psychology*, **24**, 33–44.

Butera, F. and Thurman, J.E. (1983). *Automation and Work Design*. (Amsterdam: North-Holland).

Child, J. (1984). *Organization*. (London: Harper Row).

Clegg, C.W., Kemp, N.J. and Wall, T.D. (1984). New technology, choice, control and skills. In *Readings on Cognitive Ergonomics: Mind and Computers*, edited by G.C. Van de Veer, M.J. Tauber, T.R.G. Green and P. Gorny. (Berlin: Springer-Verlag).

Clegg, C.W., Wall, T.D. and Kemp, N.J. (1987). Women on the assembly-line: a comparison of main and interactive explanations of job satisfaction, absence and mental health. *Journal of Occupational Psychology*, **60**, 273–287.

Karasek, R.A. (1979). Job demands, job decision latitude and mental strain: implications for job redesign. *Administrative Science Quarterly*, **24**, 285–308.

Martin, R. and Jackson, P.R. (1988). Matching AMT jobs to people. *Personnel Management*, December, 48–51.

Martin, R. and Wall, T.D. (1989). Attentional demand and cost responsibility as stressors in shopfloor jobs. *Academy of Management Journal*, **32**, 69–86.

Wall, T.D., Corbett, J.M., Clegg, C.W., Jackson, P.R. and Martin, R. (1990a). Advanced manufacturing technology and work design: towards a theoretical framework. *Journal of Organizational Behavior*, (in press).

Wall, T.D., Corbett, J.M., Martin, R., Clegg, C.W. and Jackson, P.R. (1990b). Advanced Manufacturing Technology, work design and performance: a change study. *Journal of Applied Psychology*, (in press).

Wall, T.D. and Martin, R. (1987). Job and work design. In: *International Review of Industrial and Organizational Psychology*, edited by C.L. Cooper and I.T. Robertson. (Chichester, U.K.: Wiley).

Warr, P.D. (1987). *Work Unemployment and Mental Health*. (Oxford: Oxford University Press).

Acknowledgements

Colleagues at the Social and Applied Psychology Unit of Sheffield University, who have contributed towards the development of this framework include Chris Clegg, Martin Corbett, Paul Jackson and Toby Wall.

Chapter 31
The effects of 'excessive' computing upon computer dependent people

Margaret A. Shotton
Institute for Occupational Ergonomics,
University of Nottingham, Nottingham, U.K.

Introduction

Much work done by ergonomists today concerns the introduction and acceptance of new technology by end-users. Special workstations and user-friendly systems are designed, health programmes developed, and new technology agreements defined in order to relieve some of the problems caused by, and the hostility directed towards, computers. In contrast, the present research has concentrated upon a group of users who appear to experience very few problems when using computers, and even find it difficult to tear themselves away from their work at the keyboards. It investigated the computer dependent person—the 'computer junkie', to use a colloquial term. The findings of the study may serve as pointers towards the features of good job design.

During the last few years a variety of academics have observed this phenomenon and expressed concern about the effects arising from computer use by computer dependent persons, effects which they felt could damage social relationships and inhibit full personal development. This research was initiated to investigate the motives and reasons behind this 'obsession' with computers, and to determine the effects arising from such an activity. The study was data-driven and evolutionary in nature, relying heavily upon the survey methods of interviews, questionnaires, scales and inventories.

Survey of computer dependents

Participants were elicited via an extensive publicity campaign in Great Britain, using the media of the national press, computer journals, and computer and radio networks. This brought forth a hundred and twenty one volunteers (the Dependents) who agreed to take part in the research. Most volunteered of their own accord, although a few were referred by members of their families.

They did not form a cross-section of the general population, but were in the main well-educated, young, adult males. Their ages ranged from 14 to 64 years, with a mean of 29.7 years, and 96 per cent of the sample were male. Of those over 21 years of age, 89 per cent had successfully gained at least 'Advanced Level' standard at school (examinations taken at the age of 18 years) and 44 per cent held university degrees, proportions significantly higher than in the general population.

All but one of the sample possessed at least one micro-computer at home (some owned four or five machines), and sixty nine used computers while at work. Table 31.1 shows the reported hours spent computing per week, both at home and at work.

Table 31.1. Reported hours spent computing per week, at home and work

Reported time spent computing	(hours per week)		
	N	Mean	S.D.
At home	120	22.4	12.3
At work	69	21.6	17.1
At home and work	68	42.8	20.4

Although Klemmer and Snyder (1972) have found that people are inaccurate when estimating the times they spend on various activities, the results showed that many of the Dependents did indeed consider that they spent long periods of time computing. One unemployed man reported spending 72 hours per week on his home micro, and over a quarter of those computing both at home and at work spent more than 60 hours per week computing. (Interviewed spouses even considered that the Dependents' estimates were too conservative and that they in fact spent more hours than they reported.)

Contrary to popular opinion the computer Dependents did not spend most of their time playing computer games. Although a few interviewees enjoyed playing elaborate adventure games, the majority spent their time programming. Most were self-taught programmers who programmed 'hands on' with little or no preplanning, and they were often familiar with many different computer languages. Their programs were ambitious, often involving months of work, and frequently exceeded the capacities of their micro-computers. This was not a deterrent, as many upgraded their equipment as soon as more powerful machines became available. Most were hardware enthusiasts, and the majority owned modems and used the networks not only to 'hack' into other systems but also to communicate with other individuals and databases.

The Dependents tended to view computers as elaborate toys, and had purchased their hardware in order to find out about new technology, not to do their household accounts. Not only were they sufficiently motivated to master new technology by experimentation, they were also prepared to invest heavily in both time and money. As the process of programming was the main function of their interaction, inertia and boredom did not occur. Computer dependency was not a short-lived phenomenon and, although the Dependents had spent an average of 5.6 years 'hooked' on computers, they still felt there were many 'thousands of man years' of interest left in the activity. As the hardware and software became more sophisticated, they continued to expand their knowledge and expertise to match the technology.

Consequences for computer design

At the start, it had been suggested that the results from this research might help computer manufacturers to design equipment which would appeal to all users and prevent the suspicion and hostility which so often occurs with naïve users. It was felt that by isolating the features of computers which held the Dependents' interest, these could be enhanced in the design of future systems so that all could experience pleasure when interacting with computers. However, the results of this study demonstrated that this was a fallacious theory.

From interviews with the Dependents and with control groups of less enamoured computer users it was possible to conclude that the features of computers which held the Dependents' enthusiasm were the same ones which caused frustration for others, because of their differences in motivation. This mirrored the result found by Malone (1984) when analysing which features of computer games made them fun to play. He concluded that the 'requirement for good toys and good tools are mostly opposite'. One's attitudes and needs determine the type of system which is required, and although the Dependents see computers as intrinsically interesting, others, who wish merely to use them as tools, tend to see them as sources of conflict rather than of pleasure.

Most users neither wish nor have the time to explore the computer at will, especially while at work, and do not rise to the challenges presented. Experimental trial-and-error methods tend to reduce efficiency and the likelihood of accuracy, and hence for most people there is a need to design straightforward computer systems which have good 'help' facilities and clear, concise and well-indexed manuals. However, unlike task-centred users of tools, the Dependents did not want the computer to be a 'black box' which could be used with little effort; they delighted in its intricacy, and were mainly unconcerned with practical end-products. They revelled in the jargon and in the fact that risk and experimentation had to be entertained in order to exploit computers to the full. They were unconcerned by the poorly written manuals, and enjoyed having to learn a 'foreign' language in order to use the machine. They had no desire to use an artifact which was 'transparent' in its simplicity, as time and efficiency were of little importance to them. Therefore, although it was possible to isolate the features of the computer and computing which had encouraged the Dependents to view the machine as a fascinating toy, such information was of little use to a computer industry wishing to design systems to encourage the frustrated or hostile user to view them more favourably.

Effects of computer dependency

The effects arising from the Dependents' computer use were also investigated in some depth, mainly via semi-structured interviews carried out with forty-five of the respondents. From the wealth of information obtained it was apparent that computing had greatly enriched the Dependents' lives (although the same could not be said of the effects upon family members).

The research indicated that for most of their lives the Dependents had experienced great difficulties with social interactions and often felt themselves to be outcasts because of their interests and personalities. The use of the computer had however given them the opportunity to reproduce an interaction which mirrored their own modes of thinking, and in addition many felt they were now able to communicate more easily and effectively with others. The use of computer networks enabled them to interact with people of like mind with whom they were able to share mutual interests, and such activity was felt to have expanded their range of friendships. This finding therefore disagrees with theories which suggest that computer dependency inhibits social development.

Their knowledge of computers had also invariably improved the Dependents' employment prospects and earnings; it had increased their prestige, confidence and self-esteem; they had gained fulfilment and a sense of achievement and they had found a relief from stress. In addition they had found an activity which was educational, intellectually stimulating, challenging and creative. One would consider that any activity which could provide all of these benefits could only prove advantageous to the individuals concerned.

Bearing in mind that the overwhelming majority of the Dependents did not wish to rid themselves of their dependency upon computers nor did they consider that they spent too much time computing, some did admit that they experienced negative effects from their activities, although these rarely outweighed the positive benefits.

Only three considered that their paid employment suffered, and this was mainly caused by lack of concentration and fatigue while at work. The fear that computer dependents waste company time while at work may have had some basis a few years ago, but with the advent of the home computer most of the Dependents worked efficiently during the day in order to return to their own machines in the evening. The studies of the school-age Dependents were more likely to suffer, but in spite of the youngsters' dedication to computing they felt confident that they would still find very worthwhile jobs within the computer industry because of the skills they had acquired. This seemed to be borne out by the high proportion of the school-age participants who had already had computer programs and articles published or who worked as contract programmers during their vacations.

Negative physical effects were few in spite of the fact that the Dependents did spend considerable periods of time in front of their VDUs, with weekend sessions of twenty-four to thirty-six hours without proper breaks being commonly reported. They often went without food for many hours, and most seemed to have adapted to needing very little sleep.

Many of the Dependents experienced adrenalin 'highs' while programming, which most found positively enjoyable. However, for one person this had led to a state of restlessness, sleeplessness and tension. His doctor recommended him to have an affair—an unusual medical prescription, one assumes!

Of the physical complaints which were reported, one person stated that his epilepsy had worsened and another that his migraines were more severe since working at the computer. However, neither were prepared to cut down the time they spent computing. Six others experienced regular visual discomfort, and seven frequently had headaches. It was of some surprise that the complaints were not more frequent considering the time spent in front of VDU screens, as any intense visual work carried out for long periods would probably induce such symptoms.

The Dependents' workstations varied considerably, from the very sophisticated to the bizarre, with bedrooms being the most common site. One young man, who complained of both headaches and visual discomfort, was unable to use his bedroom for computing as he shared this with his little brother. Instead he worked in the built-in wardrobe of his parents' bedroom. Here he worked, often for ten hours at a stretch, in a space about one metre by two metres, with no ventilation and lit only by a forty-watt light bulb. It was of some surprise that he ever survived until daybreak, and the tolerance of his parents must have been considerable.

A few people also complained of back and neck pains and 'aching bottoms', which was not surprising as many used non-adjustable dining chairs often of an inappropriate height for the keyboard. One man, who reported regular back problems, also complained of 'dents in his knees', a previously unheard of computer health hazard. His workstation was a coffee table and his seat a sofa; the 'dents' occurred when working at this low level with his elbows balanced on his knees for support. One further physical complaint came from a man who missed the sunshine and fresh air. He felt that manufacturers should produce an outdoor computer in order that he could combine his two loves, but in the meantime he stayed indoors.

Implications for job design

The reported negative effects upon the Dependents caused by their computing were, therefore, considerably less severe than one would expect to find in most commercial workplaces; they were after all voluntarily undertaking this activity for pleasure. Such results have implications for industrial management and for ergonomists aiming to alleviate the visual and musculoskeletal stresses of those working with computers. Good job design plays a vital part in avoiding or lowering the level of reported complaints. This study confirms that when one is self-motivated, working upon problems which are personally fulfilling and at a self-determined work rate, such factors remove many of the complaints frequently reported by those keying at monitored work rates, often at tasks which are personally meaningless and which lack intrinsic interest.

References

Klemmer, E.T. and Snyder, F.W. (1972). The measurement of time spent communicating. *The Journal of Communication*, **22**, 255–262

Malone, T.W. (1984). What makes computer games fun? *Byte*, **6**, 258–277

Chapter 32
Issues of new technology introduction and cross-cultural comparisons

F.W. Craven

Business Planning Director, RD Projects Ltd, London

Introduction

This paper is based on the writer's experience in introducing new technology to manufacturing industry in a number of emerging economies, principally in Latin America, the Middle East and Eastern Europe. One of the main aims in such ventures is to develop the manpower skills needed to manage and to operate a country's new and diverse industrial development projects and often to determine the role that technical education and industrial development can play in traditional societies.

Some readers of this paper may be disappointed in so far as it does not dwell on different types of new technology. To a large extent, the nature of the technology is often irrelevant—the problems stem from the newness to the recipients. The problems lie with people and consequently much of what follows is concerned with people, as managers or operators, and their training. Great differences are often found between backgrounds and practices in the Western world and in developing countries.

In many countries there is a pressing need for the speedy and economic development of efficient primary and secondary industries. Invariably this necessitates the introduction and acceptance of new technology and frequently this has to be achieved within a traditional agrarian culture. The successful introduction of new techniques and manufacturing methods depends upon the existence of people who are competent to understand and to absorb the technology and to put it to good use. Whatever the technology, its effective introduction requires the development of people who are able to manage and to operate and this must be regarded as a fundamental goal. One must not be allowed to be diverted to the more elegant, perhaps simpler and more familiar, task of producing educated people, who may be largely limited to discussion and debate. We need to develop people at all levels of management and operation who can think and act, rather than simply talk and write.

Within manufacturing technology there is a wide spectrum of concerns. Equally, when considering the introduction of new technology, there are many things to be taken into account. Two factors influenced the choice of specific matters to be dealt with in this paper, firstly, their relevance to the effective development of industry and secondly, the writer's experience in the field, particularly with regard to people. Within the wide range of topics that the introduction of new technology embraces it was considered to be most appropriate to deal with the most intractable of the problems namely:

1 The development of management and supervision
2 The training of workpeople

In such work, attempts should be made to see how developing countries can learn from the Western world, with a sensible adoption of the good and appropriate things and an intelligent appreciation and rejection of the bad things. Equally important is the need for an understanding of how certain features must be modified to suit particular and often changing circumstances. We need to identify some of the mistakes which are often made in the introduction of new technology, arising from simple errors, derogatory estimates of the competence of the recipients or, as is surprisingly frequent, from deliberately wrong—that is improper—motives. It is also important that opportunities to bypass some of the stages that the Western world has had to pass through in its more gradual development should be grasped. It is advantageous to miss out any unnecessary stages of development and to choose the most appropriate established technology, but this must be done for reasons of relevance to the industry's needs and not for reasons of fashion. Caution is necessary; the faster more advanced technology is introduced, the more exacting will the demands for thorough training become.

New technology or 'appropriate' technology?

The title of the paper is 'Issues of new technology introduction and cross-cultural comparisons'. What is meant by 'new'? New to whom? It is suggested that whereas the technology may be new to the emerging economy (indeed this will almost certainly be the case), what matters is the appropriateness of the technology to the recipients. The people who are transferring the technology must be genuine and show understanding.

Many developing countries have spent a great deal of money, both in soft and hard currency, on the purchase of a variety of manufacturing equipment, but without gaining all the benefits that they expected. In some cases this is because the suppliers have provided equipment which is out of date. Quite honestly, but mistakenly, some Western organisations have believed that the emerging nation is not yet ready for modern technology. More often, however, the motive has been to sell technology which is coming to the end of its domestic life and so to extend the return on the original investment. In other cases there has been a deliberate intent to provide old-fashioned technology, sometimes with the covert purpose of holding back the industrial development of the emerging nation. For whatever reason, much 'new' technology has been inappropriate for the industry which it has been supposed to serve and has often resulted in low levels of productivity, general dissatisfaction and mistrust. This situation has frequently been compounded by the use of outmoded technology in the local training schools.

'New technology' should mean the most appropriate modern technology which will enable the efficient adding of value to largely indigenous raw materials in order to generate income. For this to be done effectively, people have to be trained thoroughly, both in management and in a variety of trades, to meet the future demands which will arise from the development of local industries. It is of great importance that the training of people in new technology is tailored to the production of goods which are not only of reasonable quality, but also of appropriate quality. The design, quality and price must all be attractive to the changing demands of the market.

Training for new technology—an introduction

Later in the paper the relationships between managers and operators will be discussed. Before doing so, some general comments on training may be appropriate. It should be stressed that, whereas education in the broad sense is of inherent value and should be open to all, the same is not true with regard to training, or at least not to the same extent. The training, both of managers and workpeople, must be more focused. It has to be directed to the technological and other needs of specific industries, and to the effective adding of value to local materials.

Training, as opposed to education, is not an end in itself. It may seem to be stating the obvious to say that training schools should not exist as a virtual end product, but too often it would appear that this is true. Training must be directed towards the technological demands of industry, both broadly and specifically. It must not only be competent, it must also be relevant. Undoubtedly, initial training has to be basic, but it is important that the later stages deal with the specific new technology that is to be introduced. Training must be designed to suit the short term and the long term requirements of the developing industries and must be geared to retraining as further new technology is introduced. However, it has to be accepted that a perfect match between the output of training schools and the needs of industry will never be achieved.

Management and the tradesman in Great Britain

A great deal of difference has frequently been found between the methods employed and success factors when introducing new technology in countries having a long history of industrialisation (such as Great Britain), compared with some of those nations that have emerged recently and rapidly. Much of this difference stems from the contrasting relationships between management and workpeople. These relationships comprise perhaps the most important aspect of the introduction of new technology, and are worth looking at more closely.

In the Western world, procedures for the selection, training and development of workpeople and of people advancing to supervisory and more senior positions has developed in different ways over a long period of time. In the United Kingdom for example, junior and middle levels of supervision such as foremen and superintendents generally rise from the shop floor. People with experience, competence and enterprise are recognised by their management (and equally importantly by their colleagues) and given the opportunity for promotion to supervisory roles. It is not everyone who wishes to make such a move and to take on the accompanying responsibilities; indeed, people of ability often prefer to remain on the shop floor as direct workers. It is important, however, that groups of people can see that promotion is open to them and that some of their members have an opportunity for advancement, even though such openings are only taken up by a few.

The main point is that the shop floor personnel can see that there are opportunities open to them and that a majority of supervisors and managers came, and continue to come, by that route. There is therefore a collective if not a corporate feeling in the minds of many people at all levels of a manufacturing enterprise.

Secondly, there are many technical courses in the UK and other European countries which can not only add to the knowledge, skill and experience of tradespeople, but in some cases are specifically designed to support the aspiring supervisor or manager.

However, all of this has seen a dramatic change in Britain during the last few years.

There has been an enormous increase in training facilities for the service and leisure industries. At the same time there has been a lamentable lack of facilities for the training of people in conventional industry with a serious reduction in the numbers of young people taking up engineering apprenticeships. It is suggested that such failures will have a bearing on the introduction of new technology to some of Britain's workshops. Many industries and many companies have already had the experience of processes and methods becoming obsolete and the need to introduce new and more appropriate technology. There has frequently been strong resistance to the introduction of new technology which in many cases has nothing whatsoever to do with the technical difficulties of adaptation. The printing industry is a classic case.

Management and the tradesman overseas

Probably the most important thing that can be said is that in many cases, in developing countries, there is a gulf between the management and the shop floor. Reference was made earlier to a 'collective' feeling which is present in many enterprises in the United Kingdom and the reasons which gave rise to this. The understanding which comes from this can assist in the introduction of new processes and new techniques. People at various levels have at least a recognition of the problems of others, even though this may not manifest itself as a sympathy.

Unfortunately, this understanding does not always exist in enterprises which are relatively new in emerging economies. The gulf, mentioned earlier, generally arises because of the greatly superior education of many managers compared with the tradespeople, though it may be worse than this. The managers often come from a different and 'superior' stratum of society and the separation between them and the workpeople is not only immense, but is self-perpetuating. The majority of managers have obtained degrees in their own countries and many of them have taken further degrees abroad. They form an élite, and whereas, at first sight, the élite seems to be based on a superior education (which is true) the prime cause is that such people often come from different social classes from the work force. In practice, this élite is much stronger and the gap between it and the shop floor much greater in many socialist countries (which are nominally egalitarian) than we find in the Western world. This might be regarded as an unfortunate statement, but it is a fact which must be recognised and dealt with.

If we examine the history of society in the Western world, we will see that whereas it has had many upsets, there has generally been a gradual process of change. In some developing countries the change has been traumatic and it is inevitable that in the formative stages of a new, or at least different, society there will be extreme differences of class and opportunity, no matter what the genuine purpose or principles of the society may be. In the context of the subject of this paper, one result is that often a relatively small number of educated people who possess considerable authority are far removed, both in their day to day work and in their social life, from the workpeople whom they control. The difference in the automatic placing of graduates in senior positions within such societies compared with the assimilation of graduates in Western manufacturing hierarchies is worthy of reflection.

The people who are responsible for the success of manufacturing projects and for the introduction of new technology in developing countries will fail if they do not understand that the successful initiation, management and operation of such projects depends upon the full involvement of all the people concerned and upon a good balance of capabilities.

Too often the managers who are responsible for the new technology are unable or unwilling to devolve some responsibility to the workforce or their immediate superiors because they honestly but wrongly believe that such 'lesser mortals' (as they see them) do not possess the capability to carry out tasks other than the most menial ones. Frequently, of course, the motive is more selfish and often the well educated but incompetent manager retains complete authority, sometimes with disastrous consequences.

There is one positive aspect which will frequently be found when a country is moving from say, an agrarian to a more industrial economy, which is that often a wealth of innate intelligence and other talent exists in the unskilled workforce. The problems of lack of experience and training can be immense, but there is frequently no shortage of intelligent, enthusiastic people who wish to understand new (to them) technology and to benefit from the part they can play. Nowadays this is rarely true in the Western world where such capabilities are either already bespoken, or where people have become dissatisfied with life in manufacturing industry and moved to other jobs.

Training for new technology

Before commenting on training for 'new technology', an expression which is usually taken to mean 'advanced technology', a brief review will be made of training facilities for artisans, as typically found in developing countries.

A fundamental step towards successful industrialisation is the establishment of competent training schools which need to be properly equipped and managed and must concentrate on two main objectives. These are the training of workpeople in basic operations and their training towards the specific needs of industry. In some cases large factories possess their own training schools, but more often schools have been established on a national basis to deal with trainees in bulk. Basic training, while often protracted, is frequently poor because the instructors have little or no knowledge of industry and its requirements. Training schools often fail dismally in the training of people in the use of more advanced equipment. On the other hand, processes which are substantially manual and traditional, such as foundry work, sheetmetal work, welding, etc, are dealt with competently and high levels of skill are frequently developed. Too often however, trainees only become familiar with simple processes and this defect is frequently extended into the factories where the trainees eventually work. The training schools, of which their managers are often inordinately proud, can act as a virtual brake on the rate of development of a country's industries. In some cases the industrialised countries which have designed the training schools appear to have deliberately set out to provide only low technology.

It is against this rather unsatisfactory background that the introduction of new technology has to be considered. Frequently there is the all too common divide between the educated and aloof managers and their workpeople. This unsatisfactory situation is compounded by the sizeable proportion of intelligent tradespeople who are inadequately trained and frequently frustrated by poor facilities and an unsympathetic management.

There is a further problem. 'New technology' often does mean 'advanced technology'. It is frequently in the nature of advanced technology that it is also complex and its effective implementation requires the support of a variety of people of many disciplines. Of course this is not true, as far as the workman sees it, of much high volume work such as in the automotive trades. In what could be regarded as one of the early phases in industrialisation, developing countries often install bus, truck and automobile factories. Invariably these come from the Western world, and whereas some of the models may be old-fashioned, the manufacturing technology is reasonably up to date. However, the technology has almost always been proven in earlier factories abroad and little development or training is required. The operators' tasks are simple and limited and the main problems lie in material procurement and the like, not on the shopfloor.

As mentioned earlier, there are some technologies, frequently found in the small batch production fields, which require the support of several disciplines. One of these technologies is the computer numerical control of machine tools (CNC). The final section of this paper deals with the introduction of this technology to an established factory in the Middle East. All the problems that have been mentioned were encountered in this venture. At the outset, the one which appeared to be the most intractable was what was referred to earlier as the gulf between management and workpeople.

New technology and mixed disciplines

Many developing countries have introduced, or are considering introducing, computer numerically controlled machine tools (CNC) into their factories. In some cases this stems from a desire to 'leapfrog' the gradual development process that the Western world underwent and to invest in modern technology to make the most effective use of limited skilled personnel. Such machine tools are of high cost, but used properly they are extremely efficient. In many cases however, such efficiencies are not achieved.

The writer was involved with the selection and training of a mixture of managers, engineers and workpeople to support the planned installation of twelve CNC machine tools in an established metalworking factory in the Middle East. Success in such a venture could only be achieved by ensuring that everyone concerned with the machines and the process, from the management downwards (not simply the programmer and the operator) had a thorough understanding of all aspects of planning, programming, operation and maintenance. Specialised training such as this makes great demands on the trainees because of the higher levels of technology, but perhaps what is more important and certainly more difficult, it involves the training of groups which share mixed disciplines. In this particular case the group comprised managers, technicians, production engineers, operators and electrical, mechanical and hydraulic maintenance technicians.

The electrical maintenance people knew nothing of electronics, spoke little English and had no formal qualifications. The managers were electronics engineers with two degrees, fluent in English, but had no contact whatsoever with the shop floor. The operators spoke no English, had never seen a computer numerically controlled machine tool, nor indeed had they ever spoken to an electronics engineer. These were some of the real elements which had to be grasped when introducing this particular example of new technology.

The four disciplines in which people were to be trained to various degrees depending upon their ultimate job functions were:

- Machine setting and operation
- Planning and programming
- Mechanical and hydraulic maintenance
- Electrical and electronic maintenance

From the start it was evident that there would be problems due to the mixing of four disciplines with varying degrees of interaction, a mixture of qualified engineers (graduates with a different social status), technicians and machine operators, varying commands of the English language and the danger of certain aspects of the syllabus being too advanced for some trainees and perhaps too trivial for others. The setting up and the managing of this training course clearly illustrated the point brought out earlier in this paper, namely the difficulties in training people of mixed education and mixed social classes (often the same thing).

The initial training took place in the trainees' own country and consisted of three sections:

1 Preliminary introduction to Numerical Control
2 Period to allow for study of the literature
3 Further, more detailed introduction to CNC

The objectives of these pre-training elements were:

a To introduce, or re-introduce, different types of people of different disciplines and backgrounds, to the habit of training
b To introduce trainees to CNC and to the use of CNC vocabulary and terminology and to give some practice in the English language
c To provide background information showing the relevance of CNC machine tools to other types of machines
d To develop and to encourage a 'group' attitude

The remainder of the training, which was far more extensive than this pre-training, took place in England, firstly at an English language school, followed by CNC technology training at a university. This was followed by specialised training in the four disciplines in the training centre of a machine tool company. Finally, the trainees were all brought together again by industrial training on various types of CNC machine tools in a number of factories.

The pre-training was a vital element in this most successful venture. Without this preliminary training in the participants' own country, the United Kingdom content would have been less successful because the better educated people would have run on ahead and left their colleagues behind. As things turned out, when the trainees came to England the people with poorer education had already gained some grasp of the limited vocabulary necessary and the people of superior education, who initially were hesitant to speak to them, had acquired some tolerance for their colleagues. The group as a whole had an understanding of the relevance of CNC to their factory and their earlier fears of the complexity of the subject had been reduced. They already felt that, as a group, they would be able to cope.

Apart from the specialised training in the machine tool company's training centre, the group were kept together at all times. At the start of the project there were grave doubts as to whether it was possible to train effectively groups which shared such mixed disciplines and came from such different cultural backgrounds. In the first sessions in their own country, the graduate engineers and the shop floor operators were as two different races. Towards the end of the training in England it was impossible to tell them apart. The final proof was when the CNC machine tools were installed in the overseas factory. There was complete co-operation between all four disciplines and a high level of operating efficiency was achieved in a remarkably short time.

It was said at the start of this paper that there would be little comment on different types of new technology. It was emphasised that it is the people who matter and not so much the technology. This latter section dealing with the involvement of mixed disciplines in in the introduction of new technology illustrates this point. The introduction of new, advanced technology to small batch manufacture is fraught with difficulties and in this case there was the additional hazard of what was referred to as the aloof and educated manager.

Chapter 33
New technology and the quality of work: the organizational factor

B. Fruytier and K. ten Have
Institute for Social Research (IVA), Tilburg, The Netherlands.

Introduction

The introduction of computer numerical controlled metalworking machines (CNC machines) in a company may go hand in hand with major organizational changes so that the structure of the company is radically altered. In some cases, however, the acquisition of CNC machines has virtually no effect on the organization: here, the CNC machine is regarded as no more than a 'slightly more clever' piece of ordinary equipment.

The strategy chosen when the machine is introduced has effects on the continuity of the company and on the jobs of the employees concerned. This is the subject of the present article. Continuity proves to benefit from a strategy in which organizational adjustments play an important role.

The study concentrated on the use of programmable metalworking machines in machining processes. To this end we examined several company case histories. This article is based on the findings of the research carried out in two companies.

The research question

The following question was posed in the case studies:

What are the effects of the organizational structure on continuity in companies or departments where CNC machines are introduced? The following factors are considered to be relevant aspects of continuity: the cost price of the product, the flexibility of the production process, the quality of the product and the quality of work.

Design of the study

Measurement instruments

In order to chart various key data of the companies, a number of measurement instruments were constructed on the basis of existing literature, as summarised in Table 33.1. A more comprehensive explanation of the operationalization of the research variables and the approach to the study is to be found in 'CNC machine operators, production organization and flexible automation' (Fruytier *et al.*, 1988 [in Dutch]). In the present article, it is intended to concentrate primarily on the results.

Table 33.1. Summary of measurement instruments used in the study

Variables	Source
1 Company profile/market and organization structure	Intervisie/IVA
2 Involvement matrix decision-making	Intervisie/IVA
3 Work situation/job content	CCOZ-BEA (1985), Fruytier, Ter Huurne (1983)
4 Characteristics of CNC machines	Intervisie/IVA
5 Flexibility	Schepers (1984), Intervisie/IVA, Van Amelsfoort (1984)
6 Product prices	De Haan (1985), Intervisie/IVA
7 Product quality	Intervisie/IVA
8 Quality of work	M.Peeters (1985), De Sitter *et al.* (1985), IVA
9 Working conditions	CCOZ–BEA (1985), Intervisie/IVA
10 Qualification requirements	Intervisie/IVA, Bakkenist, Spits & Co. (1985)

Selecting the cases

Eight companies were selected which are pairwise comparable in terms of two production process characteristics: the complexity of the programming and the frequency of programme changes on the CNC machines. The two companies presented in this paper come into the category of high complexity of programming and high frequency of programme changes. They also largely correspond in terms of the industrial sector in which they operate and in their size.

Results and discussion

The context of the CNC machines

The key data of the companies are listed in Table 33.2. Company I pursues a policy aimed at all four aspects of continuity already defined. This company is aware that the most important improvements have to be achieved by drastic re-classifications in both the organic and the personnel organization structure. This is not the case in Company II. There, the CNC machines are regarded as no more than advanced machines. This company is attempting to tackle the changed demands from the market, which equally apply to it, by means of a more-of-the-same strategy. This difference in strategy will emerge from the following section, which concentrates on the work surrounding the machines. Where necessary reference will be made to the other context characteristics in Table 33.2.

The work surrounding the CNC machines

Figure 33.1 shows the most important tasks that have to be carried out in connection with the CNC machines. The tasks listed on the chart can be distributed among people in entirely different ways. It is conceivable that the tasks could be spread among many people and even departments, but also that they could be carried out by one person or group of people on the shop floor.

There are also organizational tasks that must be done. These might include, for example, deciding the machine/processing sequence, providing the materials, keeping a

Table 33.2. The company context of the two cases

	Company I	Company II
Size/structure	Concern 12,000 employees Company studied 150 employees Machining department 12 employees	Concern 6,000 employees Company studied 200 employees Machining department 60 employees
Market/order structure of company	-Supplies other company in concern -Machining department delivers directly to other company -Order can be processed in average of a few days -Turnover increasing as result of growth of concern	-Supplies international market -Machining department delivers directly to assembly department -Order ties up full capacity for about a year -Ever worsening position in fickle market, specifically competition from Far East
Continuity strategy of company studied		
Cost price	+ +	+ +
Flexibility	+ +	+ +
Product quality	+ +	+
Quality of work	+ +	0
Approach to change	-Integral, defunctionalization of staff departments; formation of workcells; aiming for Kan-Ban production -From two- to three-shift system	Survival strategy; it is essential to reduce cost price and above all cut delivery times. -No attempt to change traditional, rigidly compartmentalized labour structure -From two- to three-shift system, one shift to be unmanned (in due course)
Position of CNC in machining department	-New machines being installed at department at a high rate. The two processing centres studied are in one work cell -Work cells with great autonomy. Rotation of person/machine	-CNC processing centre studied is advanced compared with other CNC's in the company. It is the central machine in one of the lines formed on the basis of product and processing characteristics -Always the same person at the processing centre

+ + regarded as essential; + important; 0 not a specific area of attention

record of production data, selecting and introducing colleagues, planning days off etc. In addition to the tasks directly related to the CNC, these organizational tasks also have to be considered.

Table 33.3 shows how the CNC tasks are divided among employees in the two companies. As far as the detail in the table is concerned, it is noticeable that there are few differences between Companies I and II. The machine operator in Company I has more freedom to adapt programmes and greater responsibility for the input and output controls.

The conclusion is that the division of the tasks directly related to the work on the CNC machine differs hardly at all between the two companies. However if the organizational tasks of the companies are taken into consideration, a totally different picture arises. Table 33.4 gives an overview of the tasks carried out by the operators in the two companies in addition to their CNC tasks.

In Company I the machine operator has many non-CNC tasks, unlike his counterpart in Company II. In Company I the aim is also to transfer personnel management tasks and full production planning to the cells (see Table 33.2: the position of CNC in machining department), in addition to the organizational tasks already mentioned.

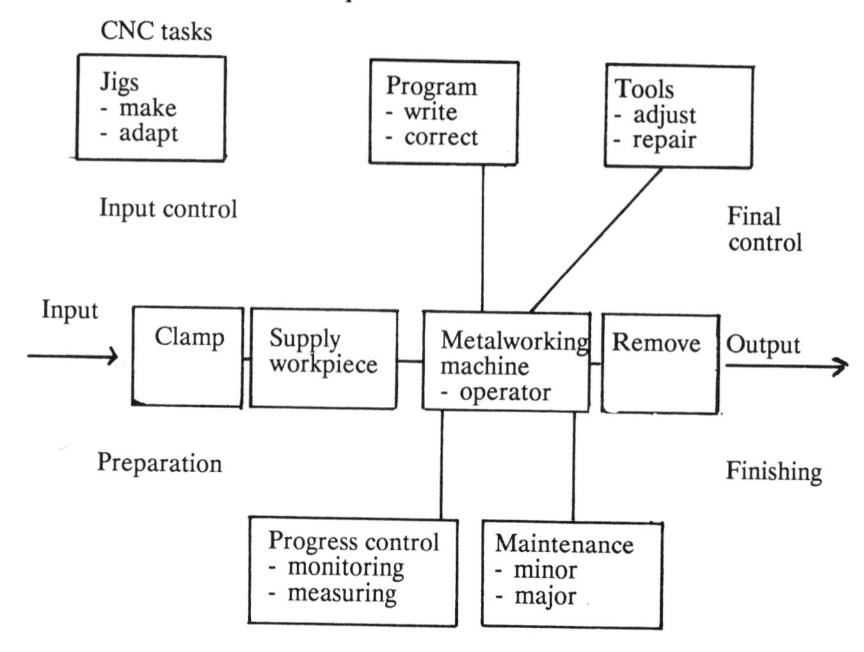

Figure 33.1. Most important tasks that have to be carried out.

Table 33.3. Division of CNC tasks

	Company I					Company II				
	1	2	3	4	5	1	2	3	4	5
Programming		X					X			
Testing programs		X					X			
Correcting programs	X						X			
Keeping up tool stockroom (machines)	X							X		
Setting tools			X					X		
Sharpening/repairing tools			X					X		
Preparation/control	X									X
Clamping workpiece	X					X				
Operating machine	X					X				
Removal	X					X				
Finishing/final control	X									X
Progress control	X					X				
Fitting jigs	X					X				
Designing jigs		X					X			
Making/repairing jigs					X					X
Minor maintenance	X					X				
Major maintenance					X					X
Machine breakdown (minor)	X					X				

1 = machine operator
2 = programmer (different department)
3 = presetter (different department)
4 = boss/charge hand
5 = others in or outside the department

Table 33.4 Organizational tasks of machine operators

	Company I	Company II
Fetching raw materials from stockroom	X	
Supplying product to next machine	X	
Deciding product sequence	X	X
Reporting quality defects and determining action	X	
If necessary, reworking product on other machine		X
Planning routing of product in department	X	
Working in newcomers	X	
Deciding (extra) days off	X	
Recording output figures	X	

The nature of the organizational structure in the two companies lies at the root of this difference. In Company I, people work in groups and the processing of an order is a group task; in other words the individual is (partly) disconnected from one specific machine or process. The organization of the work in a group is largely the job of the group itself. In Company II, on the other hand, one operator is tied to one machine, so that both the need for and the possibility of adding organizational tasks are limited.

The picture presented so far is a static one. It gives an impression of the current differences between the companies. In order to analyze the changes in the work arising out of the introduction of CNC machines (and related actions), the previous and the intended situation, must also be considered.

Some years ago, the first company strongly resembled the second. They worked in a traditional line/staff structure. The company was operating at a loss and they were faced with a choice between closing down the company or making the utmost effort to become profitable again by adopting a completely new approach. They chose the latter option. The process of change that started then is having significant effects on productivity, product quality, flexibility, and the quality of work.

In concrete terms, this means that it was decided to bring in the most advanced machines, banish the stocks from the factory, formalize quality control and the work of other staff departments and decentralize it to the shop floor. It was decided at an early stage that the basic unit of the production process should be the group, consisting of flexible machines and people who could be deployed in a range of tasks. The improvement in the quality of work is brought about by job rotation (removing the tie between individual and machine), job enlargement and job enrichment (integration of staff and organizational tasks). Although this process is by no means complete, a structure that is both productive and self-reinforcing has nevertheless been created. The group is a unit that, partly because of increased involvement, strives to expand its own functions. The integration of tools supply and, in the slightly longer term, programming, are obvious developments.

There has also been an unmistakable change in the work at the second company. The machine operator now has an easier job, specifically because the need to change tools—a major undertaking on the earlier CNC machines—has been eliminated. About every two weeks, the supervisor gives the machine operator an order, which is a list of the products to be processed and the quantities. An order usually consists of a large number of small items that do not take long to process, and a small number of large items that take about four hours to machine. An ordinary working day for a machine operator would be as follows. During the first half of the day, he machines the small items, in other words he mounts the workpieces, using the jigs, on one of the six pallets which are machined one at a time. Because the machining time is short, he is continually busy clamping the workpieces and entering the workpiece/pallet programme combination in the computer. The second half of the day is easier: a large workpiece is clamped and machined. In fact this means waiting for four hours, since things seldom go wrong with the process. Checks are made now and again, and a number of programme steps are repeated if necessary. This workplace cannot be called socially isolated, but it is functionally isolated. There is virtually no reason to look over the 'fence' around one's own workplace.

The two operators interviewed revealed very different attitudes. One, a young, relatively inexperienced metalworker, stressed the fact that the work had become easier. His colleague in the second shift, an older, experienced man, emphasized the lowering of the level of skill required on the new machine.

Effects on the continuity criteria

The main question of this study asks which changes take place in terms of the four continuity criteria (cost price of the product, product quality, the flexibility of the production process and the quality of the work) when new technologies are introduced, and the role the organizational structure plays in this. In Company I, where the introduction of new machines has gone hand in hand with an integral change in production organization, the effects of the new organizational structure and the introduction of CNC machines cannot be separated. Therefore, the overall picture of the change and its effects on the continuity criteria, must be reviewed. The results are given in Table 33.5.

Table 33.5. Effects on the continuity criteria at department or workplace level: comparison of the situations before and after the introduction of the CNC machine

	Company I		Company II	
Cost price	–dropped as a result of decreased: idle time depreciation stock costs re-setting costs labour costs		–dropped significantly as a result of lower: depreciation labour costs maintenance/breakdown costs	
Product quality	–improved, less wastage		–since fewer people work on product with fewer machines, fewer errors are made. Hence less wastage	
Flexibility				
a. product mix	+		+	
b. production volume	0		0	
c. deployment of personnel	+ +		—	
d. personnel volume	0		0	
Quality of work	trend	level	trend	level
a. required control capacity	+	high	—	low
b. internal control capacity	+	high	—	low
c. external control capacity	+	high	—	low

+ + significantly increased; + increased; 0 unchanged; — decreased

The table shows that the drop in cost price is an important consequence of the introduction of CNC machines, specifically as a result of the increase in productivity. The extent to which this cost price reduction at the machine is reflected in the cost price of the final product depends very heavily on a number of factors, including the ratio of machining time to throughput time and the relative share of the machining step in the production process as a whole. Only Company I has been able to achieve a significant reduction in stock costs by cutting down on waiting times.

As far as the quality of the product is concerned, it is clear that reducing the level of human intervention in the primary process has reduced the failure rate. The number of defective products from all the machines is so small that only marginal improvements can be made.

The possibility of making different products virtually without re-setting time has increased in the two cases, at least if the programmes already exist. If a new programme has to be written, the preparation time increases in comparison with that of conventional machines.

Only Company I has been able to prevent the flexible deployment of personnel from deteriorating as a result of the new machines, by adopting a multi-skill approach. Company II depends heavily on the individual operators. The absence of an operator can only be covered by calling in his counterpart on the other shift. The company is looking for a system with a permanent replacement.

Both companies see themselves being faced with the problem that volume flexibility will decrease when the planned increase in operating hours (see Table 33.2) is achieved. There will then be less possibility of accommodating peaks and troughs (by means of overtime).

Finally, there is the quality of work. In determining the quality of the work, only the characteristics of the work have been examined and not its evaluation by the people doing the jobs. In the definition applied, quality of work is good when there are adequate (good) possibilities of control (control capacity) in the work to meet the need for control that arises out of relatively complex tasks (required control). (For a more detailed explanation of the concept, see Den Hertog and De Sitter, 1988.)

Through the specific design of the personnel organization, Company I has succeeded in achieving improvements in the quality of the work. The drop in the number of relatively complex tasks in the primary process, because they have been taken over by the machine or by other departments, is being compensated for by adding organizational tasks to the work cell. As well as increasing the complexity, these new tasks also bring about an increase in the external control capacity. There are frequent functional contacts not only with previous and subsequent phases in the production process but also with the support and preparatory services to coordinate the work. Company II shows that using advanced machines in an organizational structure not geared to this can erode the work. The inefficiency caused by the under-utilization of the machine operator (waiting times) does not show up clearly within the company because in this workplace the machine costs are almost a factor of ten higher than the labour costs.

Conclusions

In this last section, a number of conclusions are drawn in relation to the question put forward in the introduction.

Effects on continuity

Table 33.5 gives an overview of how the two companies studied scored in terms of four continuity criteria. The two production situations studied scored positively in terms of the criteria cost price and quality of the product; no major differences were found as far as flexibility was concerned, except that the flexible deployment of personnel had significantly improved in Company I, while this form of flexibility had actually decreased in Company II. The greatest differences were found in respect of the fourth criterion: the quality of work. This improved in the first company and deteriorated in the second.

However, when looking at the figures in this table in rather more detail and setting them against the companies' needs for change shown in Table 33.2, a different picture emerges. In Company I there has been a genuine improvement in the short and long term. In this company, the introduction of CNC machines was part of an integral redesigning of the structure of the production organization.

In this way, the advantages of CNC equipment:

- reduction of the number of manufacturing phases
- free programmability within certain limits
- less product-dependence

were utilized to the utmost. Changes took place in both the nature of the orders and the processing sequence. Relatively little changed in the relationship between the company (department) and the other companies (departments) in the concern.

This change in the routine led to the implementation of new production groups and a number of the controlling tasks were assigned to it. The results was an integral improvement in production efficiency and effectiveness which, as far as one can tell to date, will also bear fruit in the longer term.

In terms of the trading-results, the introduction of CNC machines in Company II must be assessed less positively than in the case of Company I. This has to do in the first place with the company's difficult market position, but a number of internal causes can also be found. When introducing CNC machines, the company looked primarily at the cost-effectiveness of the new machines and not at the place of the machine in the organizational structure.

The productivity of the machine does indeed prove to be much higher. The labour costs are lower, as are the re-setting costs and the maintenance costs, but these beneficial characteristics of the new CNC machines have had little or no effect on the profitability of the total production process. The introduction of this machine did not solve the real problem: the company did not meet delivery deadlines. Throughput time was far too long. The reasons for this lay not so much in the nature and capabilities of the machines as in the organization of the production structure. Changes to this structure should have been made before anything else was done. The company clung to a production structure in which there were under-utilization problems in some places (workers and machines standing idle), while in other areas there were typical capacity problems (production per person and/or machine too low).

The combination of these two problems led to an increase of the throughput time in relation to the operating time, long waiting times etc. The introduction of the CNC machine did not alleviate these problems and personnel flexibility was actually reduced as a result. What this company needed first and foremost was a change in the organizational structure. It would then have been possible to make much better use of the possibilities of the CNC machines.

A number of general conclusions to sum up

1 The introduction of new technologies takes place within a specific context and often goes hand in hand with a series of changes in the organization of the production

process. The total organizational structure, within predetermined limits, must be the central subject of research (Koopman-Iwema *et al.*, 1986; De Sitter *et al.*, 1987).

2 What applies to the researchers also applies in another way to the companies themselves. The introduction of new technologies must take place within the framework of a total concept of the production organization structure in order to achieve the maximum returns. A total concept of this kind is essential if a company is to derive the maximum profit from the capabilities of new technologies, in this case CNC machines, and if the machine is not to be downgraded to nothing more than a better conventional metalworking machine.

3 When new technologies are introduced, company managements have a tendency to allow the choice of the work structure to be determined not so much by the capabilities of the machine (the departure point of the study) as by a desire to maintain the traditional structure as far as possible (Fix-Sterz *et al.*, 1986). When do companies start making the organizational structure a point of discussion? In so far as it is possible on the basis of just a few companies, some factors may be listed:
   - The disadvantages of the existing structure must be crucial and lead to major problems in the short term.
   - There must be a clear idea of an alternative structure that offers the prospect of profitability.
   - The company (concern) must have the financial resources to risk the experiment with the new structure.
   - A company will never run the risk of backing just one horse. The introduction of a new production organization structure is seen as an experiment. The possibilities of escaping back to the old situation or choosing other options must exist.
   - The personal efforts of people in the company and the powers these people have (their position of authority) determine to a significant extent whether or not fundamental changes are carried through.

As far as the quality of work is concerned, we can draw the following conclusions:

- It is possible to improve the quality of work or maintain a relatively high quality of work when CNC equipment is introduced, if the CNC equipment is slotted into a production structure based on flexible production groups (Company I).
- The quality of work in a production structure with flexible production groups can remain high because 'organizational' tasks are added to the tasks directly related to the machine, with adequate control capacity for the employees in the groups (Company I). The qualifications required for the actual work on the machines become less stringent as a result of the higher level of automation. If the job were not to be enriched by the addition of organizational tasks, the quality of work would decline. Company II is a clear example of this. The new requirements for a machine operator in this company in terms of knowledge of computer-controlled machines (his work as 'system monitor') are minimal. This knowledge can quickly be acquired on the job itself.

  This is also true of the requirements for the machine operators in Company I as far as machine-related tasks are concerned. The idea that the introduction of computer-controlled machines would lead to the upgrading of the work, irrespective of the nature of the organization, would appear to be a fiction. The reverse is more likely to be the case. A comparison with Braverman's description of the replacement of horse-drawn carriages by motorcar springs to mind (Braverman, 1984).

References

Amelsfoort, P. van (1984). Flexibiliteit: een weids begrip. *Bedrijf en Techniek*, **38**, May.

Amelsfoort, P. van (1984). Inpassen FPA in produktiesysteem. *Bedrijf en Techniek*, **38**, July.

Bakkenist, Spits and Company (1985). *Informatisering en opleidingsbehoeften*, AVO/LMBO study II.

CCOZ-BEA (1985). *BEA-voortgang*, Stichting CCOZ.

Braverman, H. (1984). *Labor and Monopoly Capital* (New York).

Fix-Sterz, J., Lay, G. and Schultz-Wild, R. (1986). *Flexibele Fertigungssysteme und Fertigungszellen*, VDJ-2, Db-128, **11**.

Fruytier, B. and Huurne, T. ter (1983). *Kwaliteit van de arbeid als meetproblem; een vergelijkende literatuurstudie*, IVA, Tilburg.

Fruytier, B., Have, K. ten (1988). *CNC operators, production organization and flexible automation* (Den Haag: COB/SER), in Dutch.

Haan, J. de (1985). Flexibele produktieautomatisering. *Research Application ZWO, FEW*, Tilburg.

Hertog, F. den and Sitter, L.U. de (1988). Integrated organization design: structural and strategical framework. Paper presented at *Conference on Technology, Organization and Job Design*, Venice, October.

Koopman-Iwema, A.M. (Ed.) (1986). *Automatiseriseren is reorganiseren* (Deventer).

Peeters, M. (1985). De kwaliteit van de arbeid in geodemokratiseerde bedrijven, unpublished thesis, THE, Eindhoven.

Schepers, H.C.M. (1984). Flexibele produktieautomatisering, unpublished thesis, THE, Eindhoven.

Sitter, L.U. de (1981). *Op weg naar nieuwe fabrieken en kantoren* (Deventer).

Sitter, L.U. de, Vermeulen, A.A.M., Amelsfoort, P. van, Geffen, L. van, Troost, P. van and Verschuur, F. (1987). *Het ontwerpen van flexibele produktie-systemen*, Socio-Technology Group, THE, Eindhoven.

Footnote

This chapter has presented some of the findings of the study *Automation and changes in the industrial workplace*, commissioned by COB/SER and carried out by Intervisie, Leiden and the IVA, Tilburg. The financial support of the Dutch programme *Technology, Work and Organization* (TAO) supported the writing of this article.

Author index

Subject index